a

D1002951

THE FIELD DESCRIPTION OF
METAMORPHIC ROCKS

Geological Society of London Handbook

HANDBOOK SERIES EDITOR–KEITH COX

The Field Description of Metamorphic Rocks

Norman Fry

Department of Geology
University of Wales
College of Cardiff

JOHN WILEY & SONS
Chichester · New York · Brisbane · Toronto · Singapore

Reprinted 1992, 1993, 1995, 1996, 1997 by
John Wiley & Sons Ltd, Baffins Lane, Chichester,
West Sussex PO19 1UD, England

First published 1984 and
reprinted 1985, 1988, 1989 by Open University Press

Other Wiley Editorial Offices

John Wiley & Sons, Inc., 605 Third Avenue,
New York, NY 10158–0012, USA

Jacaranda Wiley Ltd, G.P.O. Box 859, Brisbane,
Queensland 4001, Australia

John Wiley & Sons (Canada) Ltd, 22 Worcester Road,
Rexdale, Ontario M9W 1L1, Canada

John Wiley & Sons (SEA) Pte Ltd, 37 Jalan Pemimpin #05-04,
Block B, Union Industrial Building, Singapore 129809

British Library Cataloguing in Publication Data

Fry, Norman
    The field description of metamorphic rocks. —
    (The Geological Society of London
    Handbook Series)
    I. Title   II. Series
    552′.5   QE475.A2

ISBN 0 471 93221 3

Printed and bound in Great Britain by
Butler & Tanner Ltd, Frome and London

Contents

Acknowledgements

The preparation of this book has been assisted by many people, often unknowingly. They include my colleagues in the Geology Department of the University College of Swansea, the Editor and Authors of the Geological Society of London Handbooks, my family and friends, and those geologists working in the Western Alps during the summer of 1981 who got together and discussed geology and fieldwork on many occasions.

1

Introduction

1.1 Aim and scope

This book is about *describing metamorphic rocks and rock-masses*. It is primarily for use *in the field*, when describing those aspects of metamorphic rocks which are discernible with only basic equipment (handlens, hammer, clinometer, etc.). It has been written with final-year undergraduate students in mind, but should be helpful to any undergraduate, graduate student, practising geologist or amateur faced with making a *general description* of an area of metamorphic rocks. This book provides a systematic framework, enabling readers to produce useful and broadly similar descriptions, despite possible differences of background or specialist interest. It does not provide metamorphic specialists with assistance in the detailed interpretation of metamorphism.

This volume is a companion to handbooks on the field description of sedimentary and igneous rocks. It therefore does not cover pre-metamorphic features of obvious sedimentary or igneous origin which may sometimes be preserved in metamorphic rocks. The reader will have to decide whether to refer to this book alone or to the *handbook set*, in areas where pre-metamorphic features are preserved.

Describable features of metamorphic rock-masses may be:

1 *Pre-metamorphic* in origin (though perhaps altered beyond recognition).

2 *Metamorphic* — relating to local mineral changes.
3 *Metasomatic* — involving chemical transport and mineral change.
4 *Structural* — relating to rock deformation.

As the first three all require microscopic and chemical techniques for specialist study, there is a practical limit to their non-specialist description in the field. The limit to what may be expected in the way of structural description is less obvious. It is assumed here that production of a *map* is essential and one chapter has been given to considering the problems which can arise when mapping in metamorphic terrains. The companion handbook, Basic Geological Mapping, should be referred to for mapping techniques. *Qualitative* relationships between structural and metamorphic aspects of a rock-mass are important, and this book gives guidance on their treatment. *Quantitative geometry* and *mechanisms* of deformation are *not* dealt with, being considered beyond the scope of a non-specialist description.

1.2 Approach to metamorphic rocks

There is a widespread belief that metamorphic rocks are the most difficult rocks to understand. The techniques used in laboratory studies of metamorphic petrology can certainly seem mysterious, and they clearly do not provide any basis for describing rocks in the field. If we were to approach a rock exposure with questions such as "What is its paragenesis?", "What metamorphic facies does it belong to?", "Is it granoblastic?", we would probably come away with an empty notebook, feeling bewilderment and inadequacy. So what should we do?

First of all, we can abandon mystifying language. Questions about 'facies' urge us to jump to hypothetical interpretation when we should still be making a comprehensive description of mineral evidence. Textural terms, like 'granoblastic', urge us to put into words things which, being visual, are far better recorded as drawings. So do not be mystified by people who would have you believe that it is clever to interpret metamorphism on inadequate evidence, or to use their big words when these actually convey less information than your non-verbal records. This book will not mention 'paragenesis', 'metamorphic facies' or 'granoblastic' again. Instead, it will provide schemes for the recording of visible minerals and textures without them, and for making simple deductions about metamorphic conditions.

There are also large quantities of metamorphic rock in which minerals and textures are not visible. We have to describe what can be seen, not to worry about what cannot. If the visible features are of a different nature (veins, for example), then they are what we should describe (with the aid of the section of this book dealing with veins).

The genuine difficulty underlying work on metamorphic rocks is quite simply their variety. They cannot all be described in the same way. Thus, Chapter 6 will give alternatives for the way that composition should be described, according to whether the rock is 'fine-grained' (grains too small to be visible with a × 10 handlens) or 'coarse-grained' (grains visible). Then there are features (e.g. pseudomorphs) which may turn up occasionally in all kinds of metamorphic rock. The only advice which is always applicable is simply to ask yourself "Are pseudomorphs present?" If so, describe them (with the aid of the section on pseudomorphs). If not, move on to other things.

Clearly, this book has a duty to deal with features having as wide a variety as do metamorphic rocks. There are checklists to help a reader find the sections on particular rock features, but inevitably the reader is going to have to decide which of these sections are relevant to his rocks.

1.3 A few helpful concepts

Before explaining the structure of this handbook, there are a few matters of terminology and approach to be made clear. This book will distinguish between 'record', 'inference', 'synthesis' and 'interpretation'. A good rock-mass description does not just represent *field records*. It organizes them and edits them, making use both of direct *inferences* (e.g. that a micaschist is a metamorphosed sedimentary rock) and of *synthetic* statements. These statements of *patterns* discerned in various aspects of the rocks (composition, structure, metamorphism, etc.) together with any resulting *correlations*, make the description one of rock-mass geology, not merely of rocks. Avoid *interpretations*, that is to say, statements of processes which may 'explain' the syntheses. Remember that an excellent description is distinguished by the extent to which an over-all geometry and geological history of the rock-mass are established through correlation, without recourse to hypothetical genetic interpretations. Section 2.1 runs through the different tasks involved in making a description of a metamorphic rock-mass.

There are also distinctions to be made between different ideas of what a rock is. Ask somebody the question "What is this rock?", and you may get the perfectly valid reply "It is a milestone." As a geologist you would probably expect an answer more like "Old Red Sandstone" or "Arkosic sandstone." These examples illustrate the point that in speaking of 'a rock' we may be referring to an *object*, to a distinctive *body of rock* or to a *material*. Where necessary we can distinguish these concepts by use of the terms *rock*, *rock-unit* and *rock-type* respectively. These distinctions become particularly important in the different ways that such things are given names — a point which will be developed further in Section 2.3.

A final distinction to be mentioned at this stage is between 'outcrop' and 'exposure'. Although these words are often used as synonyms we sometimes need two definitions. The area where the solid rock's top surface cuts through a particular rock-unit can be strictly referred to as that unit's *outcrop* even if totally hidden by soil, sediment, vegetation or standing water. The parts of the outcrop which are not hidden by any overlying material are exposed, i.e. they are visible *exposures*.

1.4 Finding your way through the book

From the list of contents you will see that this text proceeds, like a description of a rock-mass, from the general to the particular. It starts with instruction on how a rock-mass may be studied (Chapters 1–3), continues by considering in turn the different field features of metamorphic rocks (Chapters 4–9), and ends with some tables and lists for immediate reference at a rock exposure (Chapter 10).

The chapter you are now reading (Chapter 1) is introductory, and with the possible exception of the Tables, it will probably only need to be read again if this book has been left unused for a time. Table 1.1 may be useful at the start of fieldwork for directing attention to chapters and sections relevant to particular types of metamorphic terrain. The introductions to particular features at the beginnings of these chapters and sections should be read at that stage. As fieldwork progresses the checklists of Chapter 10 are likely to become the more useful means of reference to particular chapters and sections, within which the phrase "*in the field*" introduces direct instructions on what to record.

Chapter 2 (entitled 'Background') deals with three different subjects, in sections which could have been presented as separate chapters. The common feature of these sections is that they are rather intellectual, and ought to be read and considered carefully. They provide, and perhaps straighten out, some ideas and concepts necessary for thorough fieldwork on metamorphic rocks. As Chapter 2 may be heavy going, avoid reading it in a rush, and use opportunities such as the journey to the field area or the time when fieldwork is halted by bad weather to read through it again.

Although Chapter 3 ('Mapping metamorphic rocks') may be treated similarly, it also contains specific advice (for example, about notebooks and photographs) of more direct use in

Table 1.1 Reference to Chapters and Sections for particular rock-mass types

Rock-mass type	Definition and description of units and banding (Chapters 3 & 4)
Slate belt	Sedimentary (and igneous) features.
Crystalline massif: igneous rocks { schists and gneisses fault/mylonite zones }	Igneous features. As below.
Regional schists and gneisses	Metamorphic rock-types and associations (plus remnants of previous structures and textures). Chapters 4–7.
Migmatites See Section 2.3.2.	Rock-type associations, including metamorphic and igneous rock-type, structure and proportion. Chapters 4–8.
Hornfelses See Section 8.1.	Pre-metamorphic features and aureole zones. Section 8.1.
Reaction-zones and metasomatic zones See Chapter 8.	Zonation, and pre-metamorphic rock-units. Chapter 8.
Fault-rocks and mylonites See Chapter 9.	Fault-rock type, properties, parentage and proportions. Section 9.2

the field, particularly during the first few days of fieldwork.

Of the Chapters on rock features, 5 and 6 deal with the core of the matter of rock description — minerals, rock-types, textures and related properties — and are relevant to all metamorphic rocks. Chapter 7 considers objects such as augen and veins which occur within some metamorphic rocks, and Chapters 8 and 9 deal with the various forms of alteration which may be associated with igneous activity, contacts or movement zones. Taken in this order, the context of each feature to be considered can be described in terms of those previously dealt with. It is therefore a suitable order in which to study and describe the aspects of a metamorphic rock-mass.

Main characteristics for the naming of principal rock-types (Section 5.3)	Textures, fabrics and fissility	Included objects
Fine-grained: physical properties (Section 5.3.3) Coarse: pre-metamorphic minerals (Section 5.3.2)	Sections 6.3–6.5.	Sections 7.1, 7.2, 7.4.
Igneous compositions.	Igneous.	Sections 7.3 & 7.4.
As below.	As below.	As below.
Metamorphic minerals (Section 5.3.1). Preserved pre-metamorphic minerals (Section 5.3.2): plus textural name for augen or for preserved textures.	Sections 6.1—6.5.	Sections 7.1–7.4.
Minerals, using either igneous or metamorphic conventions (Section 5.3.1).	Sections 6.1–6.5.	Sections 7.1–7.4.
Physical properties (Section 5.3.3): plus pre-metamorphic type, plus any metamorphic minerals visible.	Section 6.2.	Sections 7.3 & 7.4.
Metamorphic minerals (Section 5.3.1).	Sections 6.1–6.3.	—
As in Table 9.1, plus any distinct minerals, relics, physical properties.	Sections 6.1–6.5.	Sections 7.1–7.4.

1.5 Further reading

Readers may wish to undertake further reading, or to extend their study of various aspects of the geology of their field areas. To facilitate this, a number of topics are dealt with in this book in a manner which accords with particular publications listed in Section A of Table 1.2.

Further publications chosen as being of potential interest to users of this book are shown in Section B of Table 1.2. Hobbs *et al.* places a somewhat similar treatment of deformation fabrics in a detailed structural context. Readers intending to undertake full structural studies are recommended to refer to Ramsay and Huber. For the field description of igneous rocks, and structural mapping, there are the companion books of this series.

Table 1.2 Selected further reading

Section A	
Topic	*Reference*
Mapping	J.W. Barnes. *Basic Geological Mapping,* Geological Society of London Handbook, Open University Press, 1981.
Sedimentary rocks	M.F. Tucker. *The Field Description of Sedimentary Rocks,* Geological Society of London Handbook, Open University Press, 1982.
Stratigraphical procedure	C.H. Holland *et al.* 'A guide to stratigraphical procedure', *Geological Society of London Special Report,* **11,** 1978.
Metamorphic grade	H.G.F. Winkler. *Petrogenesis of Metamorphic Rocks,* fifth edition, Springer-Verlag, 1979.
Fault rocks	R.H. Sibson. 'Fault rocks and fault mechanisms', Geological Society of London, Journal, **133,** (1977), 191–213.
Pseudotachylyte	J. Grocott. 'Fracture geometry of pseudotachylyte generation zones', *Journal of Structural Geology,* **3,** (1981), 169–178.

Section B

R.S. Thorpe and G.C. Brown. *The Field Description of Igneous Rocks,* Geological Society of London Handbook, Open University Press, (1985).

B.E. Hobbs *et al. An Outline of Structural Geology,* John Wiley, 1976.

J.G. Ramsay and M. Huber. *The Techniques of Modern Structural Geology,* Academic Press, Vol.1. 1983.

McClay. *The Mapping of Geological Structures,* Geological Society of London Handbook, Open University Press, (1988).

2

Background

2.1 The stages of work

2.1.1 Preparation for fieldwork

Check:

1 That you have the necessary equipment for living, for mapping and for describing rocks safely in the field.
2 That prior arrangements have been made, where necessary, to ensure that access to the field area is physically, logistically, legally and socially possible.
3 That all base maps or aerial photographs for mapping are of suitable material, format and scale.

Read the handbooks of this series on mapping and on the description of different rocks, according to the types to be dealt with. (For many metamorphic areas, that means all of them.)

Note any conventions to be followed (arising from the definition of your task, from 'house style' or from these handbooks) regarding *rock names*, records of *orientations*, *locations* (e.g. grid references), *units* (e.g. SI only), *scales* for sketches or photographs (e.g. 1 metre, not a hammer or a lens-cap or a local coin), *map colours* and *symbols*, forms of local *place-names* (in areas of more than one language or script) and *labelling of samples*.

If you are new to this kind of rockmass description, and if there is time, try making a description of any small rock exposure showing some of the features which are specifically dealt with in this book. This may help your ability to organize time in the field by showing which tasks are easy and which are not.

2.1.2 Reconnaissance

If the size of the area and the nature of the terrain allow, walk (or drive, or ride) over the area, and try to get views of as much of it as possible from higher ground. Note where it is difficult to move around on account of the topography or vegetation, the main physical barriers (e.g. rivers, cliffs), and the main routes of access to the different parts of the area.

Locate the boundaries of the rockmass to be worked on. Note the approximate layout of outcrops of grossly different rock-types. Note, for each one, the extent and form of exposure (stream gullies, hill-top crags, river beds), and whether the good exposures display useful information (minerals, textures, structures, veins, etc.).

Note which characters are likely to define mappable units (Section 3.2.1). Note whether the size of convenient units for description is likely to coincide with that of mappable formations, or of members of them or of groups (Section 3.2.5).

List any peculiarities which may be useful but are too specific or too local for general instructions about their treatment to have been made either during geological training or in a handbook such as this. Colour, jointing, weathering style and vegetation may correlate with rock-type (but may also vary from place to place, showing dependence on topography).

Note any minerals, geometric relationships, etc. which are likely to be troublesome and try to sort out what they are. Remember when planning your field programme that they may need extra time for study.

Also note non-geological matters, such as roads, public transport, shelter, places of refreshment, etc., which can be as important as geology in determining how fieldwork is conducted.

2.1.3 Planning a timetable

A good geological description does not demand that equal time be spent on equal geographical area, or equal area of rock exposure. It may not even result from concentrating on those rocks with greatest information content. It may be necessary to spend time searching in rocks where information is scarce. Try to work out as early as possible which localities are likely to be worth spending time on, particularly for seeking out and displaying *key relationships for regional synthesis and correlation*. These may include the following.

1 Geometrical relationships, e.g. which unit cuts out which along a contact; or the direction of a metamorphic fabric in the hinge of a fold.
2 Evidence of the nature of irregularities in an over-all gradation of either composition or grade.
3 Evidence of the nature of a potential pre-metamorphic unconformity or igneous contact.
4 Evidence of either previous metamorphic (or earlier) minerals, or a previous directional fabric, in a relatively unmetamorphosed or un-deformed pod.

5 The existence of pre-metamorphic structures, such as sedimentary structures, in particular bands or beds.
6 The abruptness of a change in the mineral content between two formations.

However, do not neglect descriptions of individual rock-types. Without descriptions of all the rocks being correlated, correlation is a waste of time. Remember, also, that bedded, layered or banded sequences may contain information in the nature of the banding (cyclic variation, tectonic repetition or inversion, etc.). This may not be apparent until time has been spent, for example, in constructing a log.

So, in summary, *the tasks to be included* in a timetable are as follows.

1 To cover the field area and map it, divided into convenient sub-areas.
2 To record the particular exposures showing a lot of detail.
3 To record and display the characters of potential key localities.
4 To work out the over-all character of banded sequences.
5 To inspect closely all the contacts, boundaries and faults, and to record any spatially related features.
6 To keep some 'injury time' in reserve, for unforseen difficulties, whether geological or not.

The timetabling of these tasks will depend on the total time available, the size of the field area and the degree of exposure.

If there is the luxury of more than, say, a week to study a sub-area from the same base day by day, then for each sub-area there can be a separation by timetable of three roughly equal stages: *general mapping and recording*; *concentrated study* of complicated exposures, banded sequences and contacts noted while mapping; *recording for the purpose of display* in the final report of chosen key localities and examples of relationships discovered (logs, textures, gridded maps, etc.). Academic exercises are often of this kind.

If it is necessary, as is usually the case, to move on from one sub-area to another without visiting any exposure twice (for lack of time or because of the distances involved), then the approach must be quite different. Approximately equal time should be allocated to each formation, traverse or work sub-area (keeping some injury time in reserve) before it is known whether they are worthy of equal study. Then, within each of these, time must be allocated to the tasks listed above.

In the extreme case (which does happen) of continuous travelling in order to reach rare and very scattered exposures, then every exposure is in effect of special status, to be recorded in every detail.

2.1.4 Mapping and making field notes

Basic fieldwork consists of moving from one rock exposure to another, finding out what the rocks are and what they show. This is all that is possible at the beginning, and remains a large part of the work even towards the end. Recording is a constantly varying mixture of the following:

1 Mapping.
2 Describing and sketching details of rocks.
3 Describing and sketching details of contacts, faults and fault-zones.

Mapping demands that you trace contacts, follow marker horizons, and so on. Description requires you to record the formation that lies between such contacts and markers. These tasks are complementary. The only way to establish the uniqueness and significance of markers is by studying the formations in which they lie. On the other hand, every contact represents a relationship of some kind between the formations it separates. Ask questions about the geology of the area, and you should find that to answer them you continually shift between mapping and describing.

In noting details of rocks, it is easy to forget one kind of detail while recording another, particularly in metamorphic rocks, because they can vary in so many ways. Section 10.5 on the inside cover of this handbook is a checklist of rock features, in the order they are dealt with in this book. It may be helpful to run through it on approaching and before leaving each large exposure or group of exposures.

Ordinary fieldwork should not be continued for more than a few days at a time without *reviewing the work completed*. Such a review is needed for several reasons. You may have to pace your fieldwork to get it done. You need to acquire some appreciation of the geological picture which is emerging, and which you will have to convey in your description. You may have to make up for gaps in existing notes and organize further work.

2.1.5 Describing formations and moving on

Descriptions of individual rock units, suitable in style for inclusion in the final report, should be made as soon as possible in the field notebooks. (A separate section of notebook may be used, to avoid mixing with other notes.) These interim descriptions may need alteration at a later stage, but this can be done most conveniently *with an existing text in the field at the time that new evidence is recorded.*

There are five tasks, in addition to mapping and taking notes, which must be undertaken before you can say that the work on a particular part of the field area has been completed:

1 Description of *rock units* (including schematic sketches).
2 Description of *contacts* and *faults* (including schematic sketches).
3 Selection of *representative samples* with orientation clearly marked (to represent extremes, oddities, and *the normal rock*).
4 Checking that *realistic sketches* have been made of features which may be useful examples, illustrating points made in the descriptions.
5 Checking that all logs, etc. stated to be *representative* are really so.

2.1.6 Selecting localities for detailed work

When several units have been worked on, you should get some idea of where it may be worthwhile spending more time. This should be either to work out or to illustrate relationships which are not so clearly evident from general observation and mapping. Gridding and logging (Sections 4.2.3 and 4.1.3) are two ways of working on a different scale from that of the main map. Detailed work should be concentrated where it is likely to be geologically productive, not undertaken for its own sake.

2.1.7 Developing the over-all picture of the area

At the later stages of fieldwork, considerable thought should be given to the over-all geological framework within which the rocks will be reported. With a nearly completed field map and sections, with descriptions of formations and contacts, and with representative logs and sketches, it should be possible:

1 To put forward a *geological history* of the rock-mass.
2 To devise a *hierarchy of rock units* which reflects both real geological affinities and easy organization of material.

Look for potential alternative histories and hierarchies, in order to highlight those geological relationships which remain ambiguous. Notice any discrepancies between individual field records and the generally implied historical and geometrical relationships between units. Generally, field records must be complete, so that it is not necessary to return to previous localities to finish them. The most vital individual observations which refute an otherwise acceptable geological picture may nevertheless deserve to be confirmed at the end of fieldwork, and perhaps recorded in greater detail in order to convince readers of your description.

2.1.8 Writing up

The final description should first give a very brief over-all impression, showing the final form of correlations and hierarchies of units established at the end of the fieldwork. It should then proceed to smaller units, describing common features first, followed by individual characteristics. All else being equal, units should be described in chronological order of their formation. Table 2.1 shows the kind of framework needed.

Table 2.1 Final report: suggested layout

1 *Introductory statements:*
   The area studied, and its boundaries.
   Who did the work, and when.
   The basemaps used, their scale, source and publication dates.
   Any relevant peculiarity of the work conditions (e.g. abnormal snow-cover; the use of climbing equipment; forms of transport).

2 A brief statement of the *geographical layout* of the outcrops of different units, their relationships to topography, degrees of exposure, general weathering condition, etc.

3 Statements, preferably in the form of a table, of the *hierarchy of units* to be used for detailed descriptions, of the *definitions of formations* found on the fair-copy maps and sections, and of the *correspondence between mapped and described units*. Where stratified rock-units are concerned, an over-all stratigraphic column, drawn to scale, is a convenient way of displaying the relationship between them.

4 *Descriptions of rock-units.* Characters common to a group of units may be described first (e.g. common metamorphic grade or fabric orientation), followed by the description of units in order, subdividing as necessary. If in doubt, use the order of features in this book (banding, minerals, texture, etc.).
   Each description may be based on a detailed example map or a detailed log (chosen either as typical or illustrating extremes) together with statements of variations.
   Take care to state clearly how characters of the rock change on approaching and reaching each contact.

5 *Synthetic statements* of geometrical patterns and distributions (e.g. stratigraphic correlations made on the basis of field evidence; distributions of sedimentary facies; patterns of igneous intrusions; pattern of metamorphic grade; structural synthesis; distribution of mineralization).

6 *Geological history,* stating *what has and what has not been established.* Possible points of correlation of this history to that of other areas, or other studies.

7 *Fair-copy maps and cross-sections.*

8 *All field documents,* in the state they left the field (notebooks, field maps and sections, detailed plans and logs, etc.). These should normally be accompanied by representative *rock samples,* listed and referred to in the descriptive text.

2.2 Origins of metamorphic rock structure

If any structure can be discerned within metamorphic rocks, such structure is a variation from one point to another either in physical constitution (grain shapes and sizes) or in chemical composition. It is therefore important to have some background understanding of the possible origins of differences in composition and of the means by which these may be destroyed.

11

2.2.1 Causes of compositional complexity

1 *Pre-metamorphic complexity,* created by *sedimentary, igneous or diagenetic* processes. This can exist at all size scales. *Most compositional variation is of this type.*

2 *Veining, hydrothermal action* and *reaction-zones.*

(a) *Local veining* (segregation veining). Development of fluid-filled cracks or cavities, and their filling (as they open, or subsequently) with material from the local host rock. This may be by precipitation from solution or by crystallization of partial melt. Such processes are limited to distances of local interstitial fluid flow and produce veins which are usually millimetres or centimetres in width, although widths up to tens of metres are possible. Figures 7.16 to 7.19 illustrate veins of this type. Such veins are often the sites of deposition of pressure dissolved materials (see below).

(b) *Long-distance veins* (depositional veins). Masses of outside material, introduced and deposited along channels through the rock, as dykes, veins or pegmatites. Usually 10 cm to 10 m wide, such veins may if composite reach almost any larger size. (Veins may form by a mixture of processes (a) and (b).)

(c) *Metasomatic zones.* Successive zones of alteration of a host rock, caused by passage of extraneous fluids. They can be variable in the degree of alteration attained and thicknesses of rock affected, according to the pattern of rock permeability (interstitial or fracture).

(d) *Reaction-zones.* These are new compositions, but are dependent on existing heterogeneity for their production and their location in the rock. They are limited to diffusion distances (metres at most) from contacts.

3 *Pressure solution* and *pressure melting.* Broadly speaking, these are processes of *stress-induced, deformation-accommodated, metamorphic segregation* on a mm to dm scale (with or without input or removal of outside material). They result in directional compositional domains, related to the orientations of stress as well as strain. These processes are collectively known as 'pressure solution' to many geologists, although the dissolved material may be redeposited, and the diffusion paths involved are longer than those around individual grains during either diagenetic compaction or the deformation of a monomineralic rock.

(a) *Segregation triggered by local stress variations.* These variations occur in any rock which is either anisotropic or heterogeneous in its ductility, to the extent that it is undergoing strain. They result in local gradients in chemical potential of chemical components. Segregation is caused by material *solution, diffusion,* and *re-deposition,* giving a pattern which corresponds to that of the stress variations. This may be an individual *pressure shadow* or *local crenulation,* adjacent to more competent objects, or a *spatially regular array of crenulations,* or an *increased segregation of minerals between existing competent and incompetent bands.* The solution may be *congruent* or *incongruent.* For example, Fig. 2.1 shows congruent solution of quartz and incongruent solution of mica to kyanite in microfold limbs, with the reverse reactions in microfold hinge zones.

Fig. 2.1 Metamorphic segregation in a crenulated schist. Biotite–quartz schistosity trends from left to right. Larger pale crystals of kyanite lie along crenulations (top to bottom).

Fig. 2.2 Pressure solution stripes, causing spaced cleavage, axial planar to folds of bedding in a very low grade marble.

(b) *Segregation triggered by chemical reactions,* but spatially modified by stress variations around the reaction sites. The scope for such processes is greatest at *Very Low Grade.* They require simultaneous diagenetic or metamorphic reactions involving incongruent solution. The effects are spaced, highly directional, anastomosing compositional stripes, which do not need to conform to any pattern of material ductilities. Such *pressure solution stripes* are seen in sandstones and marbles (Figs. 7.18 and 2.2). *Stylolites* (Fig. 2.3) may be formed by the same processes at lower deviatoric stress.

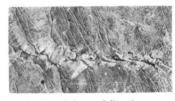

Fig. 2.3 En echelon stylolites in a very small shear-zone in very low grade marble. Extensional 'crack-seal' veinlets criss-cross the rock. They are concentrated and aligned in the zone, showing as paler streaks perpendicular to the stylolites. Field of view: 25×50 mm.

(c) *Dissolution.* Stress variation can determine preferential sites of mineral dissolution, when changes in pressure or temperature either increase the solubility of a mineral or increase the volume of solution by mineral dehydration. With subsequent removal of the solution, the over-all process corresponds to *pressure solution* in a stricter sense than does diffusive mass transfer and deposition. Geometrically, this may not be distinguishable from (a) or (b).

13

(d) *Pressure melting*. This is preferential transfer of rock material into a silicate melt according to local stress variations. It may be congruent or incongruent, and may result in precipitation, and so segregation, or in removal of the melted portion of the rock.

4 *Deformation* can enhance the previously mentioned processes of segregation, and increase geometric complexity. It can increase the reactivity of mineral grains, the permeability of the rock and diffusivities within it. It can bring reactive masses of rock closer together and extend their area of mutual contact. Tectonic repetition by thrusting or tight folding can increase the physical complexity of any pattern of compositional variability.

2.2.2 *Destruction of compositional complexity*

Chemical processes tend to convert all of a mass of rock to the same mineral assemblage, and to bring each mineral towards a uniform composition, but they do not equalize the proportions of minerals in different rocks. Their capacity to destroy geological evidence of a compositional nature is therefore severely limited. In general, once a compositional distinction exists it will be preserved, even if in a very modified form. Cataclasis may destroy the *physical* ordering of constituent parts of a rock, but short of melting, there are only two processes which can modify or destroy compositional entities.

1 *Development of reaction-zones* between rock-masses at different chemical equilibria can produce wholesale replacement of the borders of one or both rock-masses by new minerals with their own

textures. If a mass is sufficiently small, it may be entirely replaced by its reaction-zone and so made compositionally more similar to its host rock.

2 *Overgrowth by shapes of individual mineral grains.* Crystal growth normally destroys any details which existed at a finer scale. So, the growth of large grains at higher grades of metamorphism usually destroys the textures of finer grains developed at lower grade. It is aggregates of grains which define different compositional patches in a rock. These become unrecognizable if new grains grow big enough to equal earlier aggregates in size. More important in this than any general increase in grain-size with grade is the effect of strain. As soon as an aggregate patch is thinned during deformation to the point of being no wider than the individual grains of which it is composed, it will begin to break up into a string of isolated grains. These will be separated if further strain occurs, and, if the strain is sufficient, the grains of the string will become so widely spaced that the result is not distinguishable from a deformed homogeneous rock (Figs. 2.4 and 2.5).

2.3 Names and categories of rocks

Section 1.3 has already introduced a distinction between 'rock', 'rock-unit' and 'rock-type'. A *'rock'* is a real object — a boulder or an exposure — and its description would include size and shape, as well as the material it is made of. A few rocks have names as geographical features — usually things like 'Arthur's Stone', of no geological significance.

Fig. 2.4 Effects of metamorphic grain-size on preservation of a gabbro texture. Above the scale bar (only 6 cm showing) fine grain-size preserves intact flaser. Below, coarse recrystallization has broken pale streaks into separate grain shapes, and the igneous texture is destroyed.

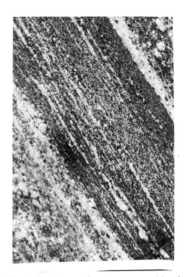

Fig. 2.5 Strings of grains: compositional patches strained to the point of losing their identity by merging into homogeneous amphibolite. Thin streaks can now only exist in the finer-grained band. Field of view: 6×9 cm.

A 'rock-unit' is a body of rock which has been designated an identity for the purpose of geological description and mapping. It consists of rocks found in spatial association in the field, and its description would include sizes, thicknesses and structures of its constituents and of their rock material. *Formations, members* and *groups* are rock-units which should be designated and named in accordance with geological conventions (see Table 2.2 and Section 3.2).

When a geologist talks of a rock, he usually describes a *generalized version of the material* of which a rock is composed. He may mention surface shapes as manifestations of the properties of the rock material, but does not normally give full details of any individual rock's external form. This is really a description of the *rock-type*, rather than the rock. Names of rock-types (e.g. 'slate', 'amphibolite') specify only the most obvious or the most important characters (as considered in more detail in Chapter 5).

A rock-type description must apply to the individual rock under consideration, but it gains its geological usefulness from recurrence elsewhere. This is so because, despite the multitude of rock characters which may exist in a rock-mass, certain characters will repeatedly occur in association (e.g. schistosity with a mica-content), such that there are only a limited and manageable number of 'rock-types' present. As geologists, we tend to take

15

this for granted. This usage of 'rock-type' is analogous to usage of 'facies' for sedimentary rocks, but in view of the heavy use and misuse of 'metamorphic facies' in another context, 'facies' is best avoided in descriptions of metamorphic rocks in the field. So we stick to 'rock-type' for the material features of a rock which are recurrent, and therefore treated as significant by geologists.

The meanings of 'rock-type', 'rock-unit' and of additional nameable classes of rocks and rock-groups derived from them are displayed in Table 2.2. Note that *rock-types* (1), *associations* (2), *formations* and *rock-units* (3) are based on visible characteristics and associations, and how these are observed in the field. Their *names* are so devised as to convey an immediate impression of an important rock character, but their full meanings in a particular rock-mass must be made known by *complete field descriptions*.

Compositional and *grade categories* (4 and 5) cannot be directly observed. They are *inferences* — simple deductions from certain visible characteristics (mainly the minerals). They are *conventional*. By specifying a predetermined scheme (e.g. Section 10.1), these categories can be named (e.g. Medium Grade, Basic Composition) without the need for further definition.

Any other kinds of *rock-type group* (6) may be invented. These may be based on *field characteristics* (e.g. 'schists'), *inferences* of common grade or pre-metamorphic character, *or both* (e.g. 'Low Grade Metasediments').

2.3.1 Which names and categories to record

At every locality recorded, the rock-type(s) should be named in the field notebook, as should the rock-type association, if one exists. These may also be noted on the field map in a highly abbreviated form (e.g. gt amte for garnet amphibolite, gr sp hfls for grey spotted hornfels).

Each *rock-type* and *association* will have to be described fully, at a stated example locality. Each *formation* will have to be described by integrating the evidence from all the localities allocated to it. *There is no need to name the formation, the metamorphic grade or compositional category in the field notes on individual localities.* Whenever the urge to do this arises, it should be channelled into making sure that *all the evidence* for such inferences has been recorded. One way to do so is to allow yourself to write such names only after you have run out of further supporting evidence to note.

Rock-type groups should have no place in basic field notes. They become important later, in the organizing of collected field evidence to make a rock-mass description.

2.3.2 Use of igneous association names

Banded gabbros, ophiolites, and migmatites are named rock-type associations which can be recognized in variously metamorphosed and deformed states.

Meta-gabbro can be an ambiguous term because of its use at different levels:

1 *Metamorphic rock-type.* 'Meta-gabbro' is a correct term for metamorphic rock-types retaining gabbroic texture.

2 *Rock-type association.* For metamorphism of a banded gabbro association, the name 'banded meta-grabbro' is preferable, because it specifies both that the

rocks have been metamorphosed and that they are being considered on a scale spanning bands of different rock-type (i.e. it is a metamorphic rock-type association).

3 *Rock-unit* (e.g. *formation*). 'Meta-gabbro unit' may be used for a rock-unit which was a 'gabbro' unit before its metamorphism. Rock-type terms generalized to the rock-unit level do not imply that the whole unit is of one rock-type. In this case, a meta-gabbro unit might include, for example, shear-zones of amphibolite and also layers of pyroxenite with no remnant gabbroic texture. It is partly to avoid ambiguity between a rock-type and a formation name that the word 'Formation' (or 'Fm') should be used in any note or reference to a named formation. The word 'Unit' can serve the same purpose in a less formal manner in the field notes about a rock-unit.

4 *Compositional category*. 'Gabbro' or 'meta-gabbro' *should not be used for compositional category,* because to do so produces unnecessary ambiguity: (= 'Basic').

Ophiolite is in practical terms an association of mainly meta-igneous rocks (in any state of deformation and metamorphism) as follows.

1 Massive ultramafic rock: rich in serpentine or olivine.
2 Banded dark meta-gabbros. Some layers may be graded. There may be some cyclical units of tens of metres thickness.
3 Massive or irregularly patchy coarse-grained pale meta-gabbros, often retaining pale igneous clino-pyroxenes.
4 Masses of patchy or streaky fine-grained, altered or metamorphosed basic rock, in which relics of pillows and their varioles are identifiable in any areas of low strain (Figs. 2.6–2.8).
5 Relatively small amounts of any of: identifiable meta-dolerite dykes; meta-chert or garnetiferous quartzite; albitite or albite-quartz 'plagio-granite'; hydrothermal carbonates; copper mineralization; red meta-sediments.

Migmatite means a mixture of rock-types, of which at least one is igneous. Several types exist. Some consist of more than one type of igneous rock. Two types are likely in areas of *High Grade* regional metamorphism, where partial melting might be expected.

1 Much granite, ramifying through biotite-rich gneisses, may be a case of a partial melt and its residue. If so, the biotitic fraction may become more depleted in quartz and feld-spars towards granite accumulations.
2 An amphibolite mass veined by a subordinate amount of acid material may represent a basic igneous body substantially 'back-intruded' by molten country rock, or inter-leaved by deformation with country rock, or perhaps both.

Fig. 2.6 Deformed pillow structures. (Scale bar: 10 cm, centre right.)

On the other hand, each of these could be a metamorphic rock intruded by later granite veins from beneath. The field evidence is vital.

For each of these associations, a full description must be made (as with any other association), but particular attention should be given to the structural relationships between the component rock-types.

Fig. 2.7 Variolitic pillow rim, cut at a glancing angle. Devitrification spherulites ('varioles') increase in size and density from the pillow edge (right) until they merge into the pillow mass (left). These igneous rock features are easily misidentified when preserved in metamorphic rocks.

Fig. 2.8 An example of metamorphic state and deformation being dependent on pre-metamorphic structure. Varioles (Fig. 2.7) have recrystallized to undeformed granular aggregates, wrapped around by a deformed matrix of chloritic fine-grained schist.

Table 2.2 Rock-types, categories, units and groups

1 ROCK-TYPE Examples: *garnet amphibolite; grey spotted hornfels.*

The constitution, texture and properties of the rock as it exists now, in its metamorphosed state, viewed at either 'hand specimen' scale of a few centimetres, or up to any larger scale to which the particular rock is homogeneous. Sections 5.4, 6.2, 6.3. *Names:* Section 5.3.

2 ROCK-TYPE ASSOCIATION Examples: *meta-flysch; migmatite; ophiolite.*

A combination of several rock-types found to occur together. The association may include particular structures, geometrical relationships between rock-types, particular order or cyclicity of any layers, and over-all proportions of the constituent rock-types. Section 2.3.2.

3 FORMATION or other ROCK-UNIT Examples: *Gleasbrook Quartzite Formation; Clegga Meta-Gabbro Formation.*

A mappable unit *(Formation)*; distinctive part *(Member)*; group of adjacent formations *(Group)*; or, more generally, any *unit* of rocks having an identity in the field. Section 3.2. *Names:* Section 3.2.6.

----

4 COMPOSITIONAL CATEGORY Examples: *ultramafic; semi-pelitic.*

A pre-metamorphic rock-type group which can be inferred from the metamorphic rock-type. Sections 5.5, 10.1.

5 METAMORPHIC GRADE CATEGORY Example: *Low Grade.*

A broad division of metamorphic grade which may be inferred from the metamorphic rock-type if it is sufficiently definitive. Sections 5.5, 10.1.

----

6 ROCK-TYPE GROUP Examples: *micaschists; gneisses; metasediments; Low Grade rocks.*

Any grouping of rock-types which a geologist may consider appropriate.

3

Mapping metamorphic rocks

3.1 Use of field maps and field notebooks

3.1.1 The field map

The field map is a working tool which *shows the status of field evidence* and *helps in the synthesis of individual records into a three dimensional picture* of the rocks being studied. *Contacts, faults* and *markers* are recorded as lines. The *outcrop areas* of different formations, and the *exposures* of these outcrops, are recorded by colour. *Locations* are marked of all key localities, of the lines of logs or transects, of any areas studied in greater detail, of peculiar rock occurrences, etc. *Field cross-sections* are attached to the field map, to avoid discrepancies between the two, and to encourage on-the-spot three-dimensional thinking.

When mapping metamorphic rocks, there is a strong tendency to record information on the map in the form of words. This is a good thing, provided it is done sensibly. Abbreviations should be used where possible, and in a consistent manner. However, consideration should be given to the use of symbols to show where certain features, such as particular minerals, have been found. Words or symbols may be marked against exposures, to indicate that a particular character has been found. Annotations indicate particular rock-types, characteristics or occurrences, *not the formation in which they lie*.

3.1.2 Structural data

Orientations of rock-bands, veins, faults and contacts, fabrics, and abstractions, such as axial planes of folds, should be marked on the map at the time of their recording, unless to do so would make them so crowded as to be unintelligible. At complicated exposures, they can be marked on a sketch map, or a sketch, in the field notebook. *Planar attitudes* of material planes (e.g. layering, schistosity), and abstractions (e.g. axial surfaces), are recorded as *strike* and *dip*. *Linear directions* of both linear elements (e.g. fold hinges) and intersections of planar features (e.g. bedding/cleavage intersection) can be recorded as either *bearing and plunge*, or as *pitch* on a measured plane. In every case it must be clear what the reading is of (see Chapter 5 of Barnes 1981).

Conventional symbols exist only for bedding and banding, fabric planes, joints and minor folds. All other measurements must be clearly annotated. Symbols should be marked on the map in the field, in order to show up any discontinuities in directions while these can be checked (to see either to what geological feature they correspond, or that an erroneous record has been made). Similarly, minor fold sense should be recorded immediately, so that it is realized from a reversal of sense on field map and section, that the sense of traverse of a layered sequence may be reversed by a major fold closure. This may be a key locality for establishing the relationship between layering and fabric.

The cross-section attached to the field map should show all contacts, representative traces of any banding and fabrics, traces of axial surfaces of upright folds and the senses and style of minor folds.

3.1.3   Notebooks

The *rock-types* should be recorded at each locality, as should the *association* where appropriate. It is wrong for notes to state only the formation to which rocks are allocated, or their metamorphic grade, or their compositional class. *Field notes must first record the evidence* on which such generalizations are based, not the deductions themselves (Section 2.3).

Whenever possible, descriptions of rock features should be in the form of, or as annotations to, a *field sketch*. Sketches record a lot of information which no one would think to put into words, even if they knew how.

Orientation data should be recorded in the field notebook. This includes duplicating readings recorded on the field map. If the map and notebook records of the same reading are later found to be contradictory, they should be deleted. (This happens surprisingly often.)

Attempts at extrapolation of igneous, sedimentary or structural patterns (e.g. *trial regional cross-sections*) should be made in the notebook for direct comparison with those attached to the field map and with what is found next in the field. It is a good idea to have hypotheses which can be refuted either by inspection of previous notes or by further work.

Interim descriptions of formations should be in field notebooks, so that they are available for instant reference at any later time in the field. That way, minor changes which take account of new evidence will be made, not forgotten.

3.1.4   Photographs

Geological photographs should include a universal standard of scale (1 metre, 1 cm, etc.). All features photographed in the field should be sketched as well, as a note to remind the photographer of what the photograph was supposed to show. As with other sketches, the attitude (strike, dip) of the rock surface and the direction (pitch) of lines on it should be recorded. For wider views, a record should be made of the viewing direction. The viewpoint for all photographs should be marked on the field map. Remember that a drawing which has been made for the purpose of displaying a particular feature or relationship will usually do so better than a photograph. Therefore, photographs should supplement drawn illustrations, not the other way round.

3.2   Defining and mapping formations and markers

3.2.1   Formations

Production of good maps and field sections is essential to the description of a rock mass. This involves, fundamentally, dividing the rock-mass into 'formations' (mappable units) with precisely specified diagnostic features, and displaying their geometrical pattern on maps and sections.

Formation definitions and distinctions may be more complicted for metamorphic rocks than for sedimentary or igneous ones. A basically one-dimensional 'stratigraphic' or 'time-dimensional' approach will normally work well for sedimentary or igneous rocks. This identifies characteristics, perhaps locally variable, of rocks formed during successive episodes of geological history. The over-all history of a metamorphic rock is usually more complicated, consisting of a pre-metamorphic history, plus one or more metamorphic histories, and perhaps also deformational histories, with the deformation and metamorphism not only overlapping, but varying in intensity and style with position as well as with time. Division of the rocks may be independently possible for each of these histories, and on the basis of intensities and styles.

It is necessary to establish a clear *hierarchy of distinctions* whenever more than one kind of criterion is used to characterize formations. For example, a distinction may be made at group level between a meta-sedimentary group and a meta-igneous group. At formation level, the meta-sediments may be divided by metamorphic grade, and the meta-igneous rocks by division (of significance yet to be established) between 'massive', 'banded' and 'banded-and-strongly-veined' formations. For mapping at formation level, this will mean that each formation is defined by a combination of characters, and different criteria may characterize different stretches of a formation boundary. This aspect should be carefully checked whenever boundaries are followed.

In general, the use of the same distinguishing criterion at different hierarchical levels within different groups should be avoided. It is perfectly permissible for a group to consist of only one formation where necessary. Possible kinds of criteria for distinguishing formations include:

1. Pre-metamorphic rock-types, including igneous and sedimentary structures:

2. Metamorphic grade, either by index mineral presence/absence or by change of whole mineral assemblages:
3. Existence or intensity of deformation or resulting fabrics, or differences of structural style:
4. Metamorphic grain size. (General grain size, or whether particular minerals occur as porphyroblasts.)
5. Colour:
6. Style of weathering or fracture:
7. Degree of homogeneity/heterogeneity (banded, massive, patchy):
8. Presence or absence of veins, pegmatites, dykes, pods, accessory minerals, etc.

Apart from the possibility of distinguishing formations by several criteria of different kinds, the general approach to mapping formations is the same for metamorphic rocks as for others. It is necessary to specify defining criteria and to stick to them. After mapping, information collected on trends and extents of lithological variation within a formation can be used for the description of the formation. The specifications of the defining criteria may need slight amendment as a result of experience during the mapping, but it is important that none of the criteria should break down when faced with what the rocks are really like. If a definition used to map a small part of the area becomes unsustainable, it is best to re-map the area. (Notebook records should not need repeating.) Otherwise the area mapped on different criteria from the rest must be described and reported as such at the end of the work.

3.2.2 Markers

The geometric pattern of rocks, at a suitable scale for representation on the map, is constructed by tracing out 'markers'. These may be unusual rock layers. Such *marker bands* may be sufficiently thick that their boundaries can be accurately represented on the map. Otherwise they must be marked conventionally by a double line. The marker may be a sudden change in rock-type, the plane separating two rocks being a *marker horizon*. For practical purposes, any unusual gradational change sharp enough to take place over a distance too small to be representable on the map, or any very thin marker band, can be used as a marker horizon. Even with a change taking place gradually over a great distance, it is possible to specify a point on that gradation, and to trace its locus as a marker across an outcrop. Markers should be followed in the field and drawn onto the map whenever possible.

There is a direct relationship between markers and formations. To trace a contact between formations is to use the contact as a marker. Any distinctive marker is a candidate for a formation boundary, unless this would produce unsuitably small formations. Any large outcrop may be subdivided into smaller formations along suitably located marker bands, even if no clear difference can be seen between rock-types on the two sides. If there is a gradual change between two formations, their mutual contact is best defined to be at a marker band, if one exists.

3.2.3 Stratified sequences

The concept of markers derives from, but is not restricted to, originally stratified rocks. If layered rocks exist in the area, it is likely that some division of the sequence into formations, and identification of marker bands, may be needed. This should be done with full awareness of potential structural complications of thrust repetitions, inversions on folds, etc. Two similar bands may well turn out to be parts of the same original layer. Any unusual but recurring rock-type should be noted early on. If layered sequences occur only at two widely separated outcrops, there is likely to be a problem in deciding whether they are the same formation or not. The same pattern of markers may well be the evidence that correlates them, and this possibility should be looked for. Possible contrary evidence, such as non-equivalence of markers, or large differences in metamorphic grade, should be noted at the same time. In general, it is good practice to tackle banded sequences early on, as they can show both features and difficulties which will help in later handling of outcrops of less informative rocks. However, the physical anisotropy of a layered sequence makes it liable to structural complication during deformation, particularly to folding and to imbrication, which have no cause to develop in homogeneous massive formations. Otherwise it is good practice to map less deformed or less structurally complex formations first. Where the pre-metamorphic state consists of a stratified sequence of sediments, or lavas, or both, unconformably overlying a previously metamorphosed and deformed basement, then the two approaches (to map the stratified rocks, and to map the rocks which are structurally less complex) will coincide.

3.2.4 Scattered blocks, dykes, veins, pegmatites, etc.

The method of treating scattered blocks or sheets of rock which are different from most of the formation in which they lie will depend on their size, abundance and compositional consistency. There may, for example, be a set of veins which, by similarity of orientation, shape, size and composition, can be assumed to be related. If these are numerous, they should be included as an essential component of the formation or member in which they lie. (Therefore, outcrops of similar host rock which lack veins belong to another formation or member.) If veins are few and scattered, they constitute a 'member', and should be recorded either accurately or symbolically on the map, according to size. If large enough, veins may be considered formations in their own right, particularly if they cross-cut several formations of host rock. However, for description, different parts of the vein should then be considered separately because of the importance of wall-rock alteration, reactions, and local changes in vein minerals, dependent on the particular host rock. The same considerations apply to pegmatites, dykes, and scattered undeformed pods in deformed formations.

3.2.5 Units for description

The units of a rock most suitable for description may not coincide in scale with the mappable formations. They may be either larger or smaller, but the relationship in each case must be clearly stated. The size of formations may be limited by map scale, and that of descriptive units by available field time or report length. These limits should be consciously determined. One of the causes of indecision in the field is being unclear where to generalize and where not to. The best approach in the field is to record fine details of one or more chosen examples of each rock unit (their compositions, textures, type

sections, complete type log, plans of type localities, etc., as appropriate), and then to note the variations from these through the unit. The number of units may be restricted by the time available. If the examples of several field units show the same features (e.g. mineral contents), the report may combine them for description of the particular aspects of the rocks concerned. Any additional generalizations will be due to restrictions on report length rather than to geology.

3.2.6 Naming formations

There is no reason why the names you give to formations of metamorphic rocks should not accord with standard procedure. They should consist of three parts: *Name, Rock-type, Formation* (e.g. Gleasbrook Quartzite Formation).

The *Name* should be the *proper name of a geographical feature,* such as a habitation, mountain or river, in the vicinity of the formation's type area. It must be distinctive, so avoid names commonly given to geographical features, such as (1) parochial saint's names of towns and villages, (2) words having a meaning other than as proper names (e.g. Ford, Bridge), (3) common figurative names (e.g. mountains called 'Sugarloaf', 'Pain de Sucre', etc.), including derivations from local or ancient languages (e.g. rivers called Avon, Stura, Rio, etc.). Avoid any name already in use for a lithostratigraphical unit.

The *Rock-type* term should normally indicate *the most common rock-type* within the formation. *It does not imply that the whole formation is of one rock-type.* It should be as specific as is reasonable. For example, a formation of quartzitic meta-sandstones interbedded with meta-arkoses, marbles, micaschists and calc-silicates, is better indicated by the rock-type name 'quartzite' than by the more universally correct rock-type group 'meta-sediment'.

In other cases, this term signifies not an individual rock-type, but a *rock-type association* (e.g. meta-flysch, banded meta-gabbro). (See Table 2.2.)

Avoid rock-type names which are local (e.g. killas), or imply significance of a geotectonic or genetic character (e.g. supracrustal), or those without agreed limits to their meaning (e.g. epidiorite).

The word *'Formation'* makes it clear that it is a rock-unit, not a rock-type, which is being referred to.

When describing an area which has been mapped before, do not use the previous formation names unless you are also accepting the previous definitions and the previous mapping. This book addresses itself to the task of a new rock-mass description. It may be that, once this has been completed, there will be a need for comparison with previous work, and perhaps the need for an agreed redefinition of units. That is a separate (and later) matter from producing a good description which can be seen to justify such a comparison.

3.3 Contacts and boundaries of metamorphic rocks

3.3.1 The nature of particular contacts

Contacts of metamorphic rocks can be ambiguous and very difficult to deal with. They may be pre-metamorphic (depositional, intrusive, or faulted), or may represent a change, either discrete or gradational, produced during metamorphism (metamorphic grade, metasomatism, abundance of minor

intrusions, finite strain, intensity of deformation fabric, structural style or compositional variability). They can be not just of many different types, but of several types at once. For example, original compositional difference may determine that one rock unit has new metamorphic minerals and has suffered intense strain while a neighbouring unit has not. In deformed rocks, any of these types may since have been geometrically modified, so destroying or hiding previous angular discordances or other geometrical relationships.

Yet, as in other rocks, each contact is in a sense a combination of marker horizon and key locality at which some relationship between rock units exists. It is because there are so many possibilities in metamorphic rocks, that it is so vital to pay attention to the contacts among them. Also, the boundaries of a metamorphic rock-mass with non-metamorphic rocks should only be interpreted after a full appreciation has been gained of the nature of the metamorphic rock-mass's internal contacts.

It should be standard practice to walk along contacts, and to look to some distance to each side, on the alert for:

1 Angular discordances which turn into the line of the contact:
2 Changes in strain or in metasomatic alteration or in retrograde metamorphism towards the contact:
3 Faults belonging to the same system as that of the local contact, to which the main displacement or contact may transfer further along its length:
4 Any features which have been offset, either by a straightforward fault, or by a contact zone which (though perhaps originally defined in some other way) has become a zone of ductile displacement.

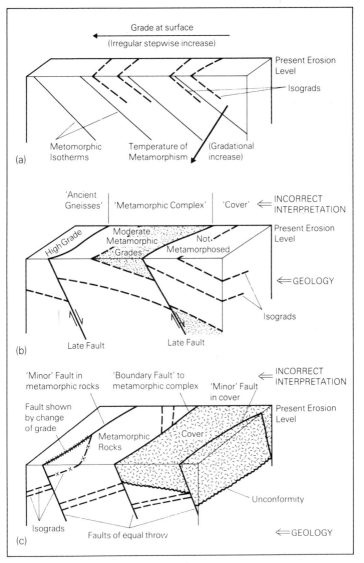

Fig. 3.1 Faulting of metamorphic boundaries. (a) Grade can increase in sudden steps despite gradational changes of metamorphic environment. (b) Incorrect interpretation of time relations across late faults. (c) Interpreted fault significance a fortuitous result of erosion level.

27

3.3.2 The regional significance of major boundaries

It may be that the natures of the contacts in a field area are locally unambiguous. There may still be ambiguities about their wider regional significance. In this respect it is a mistake to prefer sharp distinctions between rock units, because they seem simpler, when many fundamental geological distinctions are more likely to be gradational. In this context there is an apparently common prejudice against accepting gradational passages either between rocks of moderate grade and rocks which are almost unmetamorphosed, or between rocks of moderate grade and various kinds of gneisses. It is good to be on guard against such prejudices.

Figure 3.1 illustrates some possible relationships in which the wider significance of boundaries is easily confused. These show how easy it is to overplay the importance of faulting and to underplay metamorphic gradational passage.

Because the boundaries of the metamorphic rock-mass against unmetamorphosed rocks represent some regional feature, it is worth having a very good think about the nature of these boundaries on a regional scale. *Is the character of the contact at the local scale really representative of the regional character?*

4

Banding

This chapter is about 'banded rocks'. These consist of a stack of compositionally definable rock layers of sheet or pancake-like form and having a banded appearance on cross-cutting exposure surfaces. If such rocks are encountered, an attempt should be made to characterize the banding pattern. The individual rock-types of the different bands can be described afterwards in terms of their mineral contents and textures. The modes of occurrence of compositional bands are varied, and Table 4.1 suggests how different types of bands may be initially distinguished for purposes of rock-mass description. See also Fig. 4.1.

4.1 Gross banding

4.1.1 The nature of gross banding

This section concerns any banding at a scale of one metre or more which is a general feature of a formation, and which cannot be characterized completely in igneous or sedimentary terms. Most bands of this size have a pre-metamorphic origin and provide information about the pre-metamorphic geology. The exceptions to this are possible veins or igneous dykes and sills intruded during the metamorphic history. Smaller scale compositional bands, at sizes of one metre down to

Fig. 4.1 Banded metamorphic rocks. Gross banding is shown by the pale band (left) and more variable darker bands (extreme left and centre right). The pale band can be subdivided into three portions, each with its own type of fine banding. Width of field of view: 4 m.

Table 4.1 Procedure for banded rocks

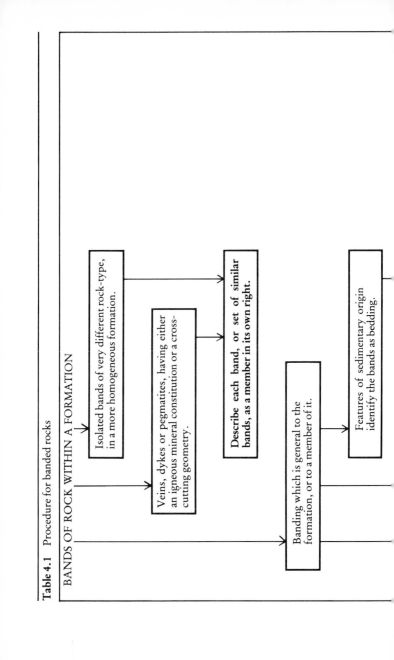

BANDS OF ROCK WITHIN A FORMATION

Isolated bands of very different rock-type, in a more homogeneous formation.

Veins, dykes or pegmatites, having either an igneous mineral constitution or a cross-cutting geometry.

Describe each band, or set of similar bands, as a member in its own right.

Banding which is general to the formation, or to a member of it.

Features of sedimentary origin identify the bands as bedding.

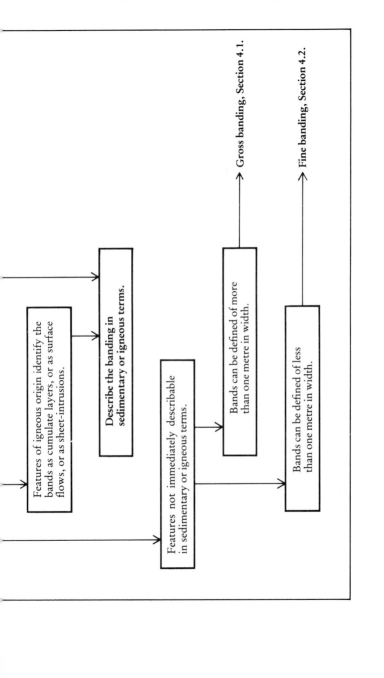

Features of igneous origin identify the bands as cumulate layers, or as surface flows, or as sheet-intrusions.

Describe the banding in sedimentary or igneous terms.

Features not immediately describable in sedimentary or igneous terms.

Bands can be defined of more than one metre in width. → Gross banding, Section 4.1.

Bands can be defined of less than one metre in width. → Fine banding, Section 4.2.

individual mineral grains (Section 4.2), may be superimposed on the gross banding.

Although the sheet-like three-dimensional forms of rock 'bands' may have existed in an original sequence, either of beds, or of parallel intrusive sheets, or of igneous layers, substantial alteration of the character of such sequences is normally accomplished only when considerable deformation accompanies metamorphism. If that has been the case, there may be some bands which originated as cross-cutting intrusions, but have had their angular discordances so reduced by high strain as to become indiscernible. Other bands may have been massive equidimensional xenoliths or olistoliths, which have been given pancake-like forms of considerable lateral extent by high strains. In whatever form, a gross compositional layer still represents a compositional entity of pre-metamorphic or intrusive, rather than metamorphic, origin. It is important to record (1) the existence of banding, (2) any evidence of the original nature of either the whole sequence or of components of it, and (3) evidence of metamorphism, or of deformation, or both.

4.1.2 Questions about banding

Is banding discernible? Is it obvious, with bands having grossly different colour or mineral content? Is a more subtle banding present elsewhere, consisting of slighter variations in mineral proportions? Are there bands with different resistance to weathering, or different weathered colour, which may not otherwise be distinguishable? Is there variability from place to place in accessory mineral content, or in grain size, or in degree of grain aggregation, which may on careful examination turn out to be defining parallel bands of more constant character?

What features constitute and characterize the banding? Do bands differ (1) in that there are different minerals present? or (2) in mineral proportions? or (3) in grain sizes? or (4) in grain shapes? or (5) in degree of grain aggregation?

Do discrete bands of homogeneous character exist, or do the rocks constantly change all the way through the sequence? Are certain compositional bands characterized by sharp boundaries, and others by gradational boundary zones?

As the sequence is traversed, is there a shift (1) in the preponderance of certain band compositions, or (2) in the relative thicknesses of different band-types, or (3) in absolute band thicknesses?

Do some bands show boudinage or 'pinch-and-swell' (Fig. 4.2)? If so, is this feature common to all bands of the particular composition, or does it exist only where neighbouring bands are of a particular compositional type?

4.1.3 Graphic logs

A graphic log can be a convenient way of portraying field data from banded metamorphic rocks. Such a log is similar to that for a sedimentary sequence, but there are no existing conventions on how the information should be portrayed. It should include the three main columns used in Fig. 4.3. The first column is graphic. It includes 'thumb-nail' sketch caricatures of small structures, and could also have symbols marking particular mineral occurrences. The second column is a continuous trace representing the principal variable character (darkness in this example). The third column is for brief written notes. The

Fig. 4.2 Pinch-and-swell of the darker layers in some deformed banded rocks. (Scale bar, top left.)

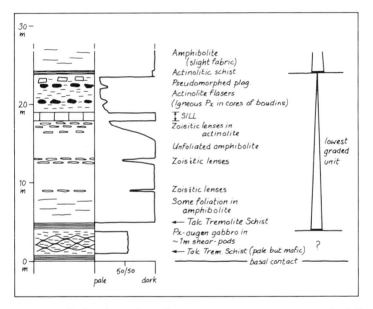

Fig. 4.3 Graphic log of part of a banded meta-gabbro sequence as recorded in the field notebook. See text.

log illustrated in Fig. 4.3 was constructed to elucidate grading, so a fourth column has been added to show the graded units identified.

Here are four cases where logs have proved useful.

1 Meta-sedimentary schists were logged in adjacent outcrops in an attempt to establish a common lithostratigraphic succession, and for correlation between areas.

2 Meta-gabbro schists were logged to portray cyclical compositional variations and grading at a scale of

33

tens of metres in an attempt to distinguish original cyclicity from tectonic repetitions within the sequence.

3 A deformed greenschist sequence, seen to exist in places of previous sub-parallel sheet intrusions, was logged to discover whether there was symmetry from side to side of the sheets, as might be expected from a sub-vertical 'sheeted dyke complex', or whether a consistent asymmetry existed which might be interpreted as indicating a 'way up' of a set of sub-horizontal sills or lava flows.

4 Gneisses consisting mainly of alternating sheets of amphibolite and granite were logged to display graphically the manner in which the granite sheets became gradationally both more abundant and larger towards an area of granite gneisses, and to search for any accompanying changes in other rock-types and in tectonic structures.

Although symbols and abbreviations are used on logs as a way to portray information in a succinct form, a graphic log is of greatest help for those continuously variable rock characters which may be represented by a column of variable width. Clast size is the chief such character of clastic sedimentary rocks. Grain size may be the character in some metamorphic rocks, whether of pre-metamorphic grains (clasts or igneous minerals), or of metamorphic minerals. The character in the examples cited above was (1) carbonate mineral content, (2) colour (darkness), (3) metamorphic grain size, (4) whether granitic or not. In all but the last of these (a simple yes/no case), chips of rock, representative of various degrees of the variable concerned, were carried in the field for comparison. This is particularly important in trying to bring objectivity into such things as colour.

Differences in the representative nature of the logs of metamorphic and sedimentary rocks can be important. In sedimentary rocks the sub-horizontal depositional attitude of bedding ensures considerable continuity along the layers. In metamorphic rocks such continuity cannot be assumed. Successive low-angle cross-cutting relations are possible, and the extent of lateral continuity must be established, either directly by mapping, or indirectly by reference to a depositional igneous or sedimentary origin. Even then, it is usually necessary to establish the extent of lateral variations in thicknesses caused by deformation.

4.1.4 Way up, repetition and cyclicity

Whether by graphic log or otherwise, repetition, cyclicity, grading and any generally asymmetric features should be looked for, both at the scale of the banding and at a scale which spans many bands. Some bands may be grain-size graded, while others may become finer (or coarser) equally towards the two sides of the band. Some may be graded in their mineral content (Fig. 4.4), or in their refraction of a cleavage or schistosity (Section 6.4; Figs. 6.14 & 6.15). Some may have been deformed into cuspate shapes on one side only. If any such asymmetry exists, the sequence probably originated as a sequence of rock layers, and the asymmetry probably represents indirectly a 'way up' (i.e. the asymmetry between upwards or downwards in originally sub-horizontal layering). Such asymmetry is unlikely in sequences derived by intense deformation from more equidimensional shapes, or in vertical layers, or in the products of chemical transport at great depths in the crust.

Fig. 4.4 Grading of mineral content in meta-sedimentary schists. Field of view: 5.5×8 cm.

Questions about asymmetry. Does any asymmetry exist? Is it characteristic of certain compositional layers only? Are there reversals of asymmetry, indicative perhaps of early isoclinal folding? Can any such potential fold reversals be corroborated by more direct evidence, like intrafolial folds, or reversal of sense of slight discordances between mineral fabric and banding, which may only be visible in the potential hinge zone located by the reversals of banding asymmetry?

Questions about repetition. Is there repetition of sub-sequences of several bands within the over-all sequence? How precise is this repetition? Is it never quite the same, suggestive of oscillatory or cyclic processes of sedimentation or magma settling? Does the nature of cycles gradually change through the sequence? Is there nearly exact repetition of minor details belonging to one sub-sequence, or a sudden step backwards in an over-all change? These suggest tectonic repetition. If so, do the rocks at the break show any sign of isoclinal folding, rather than a clean thrust?

4.2 Fine banding and striping

4.2.1 Differences in scale

This section deals with compositional layering, or discontinuous streaking or striping, on scales between those of individual grains and of about a metre wide. Although many of the questions in the previous section can also be asked about banding on this finer scale, the answers may have quite different significance. These size-scales of banding differ (1) in their potential origins, (2) in the methods to be used for their observation and recording, and (3) in their uses to both authors and readers of rock-mass descriptions. The choice of one metre as the dividing line is arbitrary, and may be changed if, for example, 0.2 m or 2 m visibly corresponds to a qualitative change in the nature of the rocks being described (Fig. 4.1).

In terms of chemical compositions, bands more than a metre wide, even if somewhat changed during metamorphism, were probably either distinct entities before metamorphism, or were intruded during metamorphism. At scales up to perhaps a metre, and certainly at less, chemical diffusion commonly operates so intensely during metamorphism as to cause wholesale chemical changes where adjacent bands are not in chemical equilibrium. At each stage of metamorphism, rock compositions are altered by diffusion in such a way as to make more similar the mineral combinations of which neighbouring bands are composed. Adjacent compositions developed during one episode of metamorphism may be thrown out of mutual equilibrium by later conditions, leading to further diffusive transfer. Such processes do not at any stage work to equalize proportions of minerals, but alter proportions in complicated ways to produce convergence of mineral assemblages. Some compositional bands will be enlarged by diffusive influxes, others diminished. New bands will form as reaction-zones at contacts. Others will be used up, or have their distinctive character completely destroyed. There are also several processes of stress-induced chemical segregation (Section 2.2.1(3)), which act on the scale at which considerable diffusion is possible, and may alter the nature of banding.

Although it is, above all, the possible chemical origins which require a separate approach to the interpretation of bands at smaller scales, there are other complications. The variety and abundance of pre-metamorphic patchiness are very great at small scales (bedding lamination and cross lamination, flow banding, trace and body fossils, igneous layering, and deformed and streaked out amygdales, vugs, clasts and phenocrysts). In addition, larger original features such as metre-scale grading in beds or igneous layers, may be reduced in thickness by tectonic strains to this scale.

As implied in Section 4.1, a prime task when recording gross banding is to span many bands, and to recognize patterns in their sequence. The gross bands are easily visible individually, but data on them must be collated (e.g. in a log) to reveal larger patterns of which single bands are mere components. When recording finer scale banding, it is the patterns which are easily visible, and effort is needed to specify both the different component bands and their relationship to each other.

Recognized patterns of gross banding, together with reasonable assumptions about pre-metamorphic or intrusive origins, often permit simple deductions to be made about the nature of banded formations, and these may be used in producing the rock-mass description. At the finer scale, complications normally require specialist study and a hypothetical interpretation before any further deductions can be made. So, in a non-specialist description the task is simply to make accurate and detailed accounts of well chosen and accurately located examples of fine banding.

4.2.2 The nature of fine banding

Examples chosen for description should illustrate the general nature of fine banding in the formation, extremes of its normal variation, and also cases where its nature is distinctly different from the general case. These can only be chosen when the general nature of the banding has been established.

Questions about fine banding. Are there continuous thin compositional bands, or discontinuous lensoid stripes and streaks, or both?

Are these found throughout the sequence, restricted to within gross bands of certain compositional type, or found everywhere except in gross bands of a particular type?

Do individual bands or lenses have sharply defined edges, gradational edge-zones, or are they almost indefinable, representing a general gradational patchiness of composition?

Do bands thin laterally, and then break up into small boudins or lenses? Do they thin to a discontinuous string of individual mineral grains (as in Figs 2.4 & 2.5)?

Is there any apparent relationship between band thickness and composition? Do some bands get broken by cross-veins, and not others (Fig. 4.5)? How much is banding due to differences in proportions of the same set of minerals, and how much to changes from one set of minerals to another? What are the normal proportions of different band types? Do these proportions vary either locally, or across a formation? Does the fine banding change according to the gross banding pattern, where one exists?

Fig. 4.5 Cross-veins in greenschists, showing up bands on a different scale from those immediately apparent. Field of view: 25×33 cm.

4.2.3 Gridded maps of example localities

It is likely that an example locality can best be described with the aid of a large scale map or plan. This can be constructed accurately on squared paper by using a rope or string grid, laid over any relatively flat exposure of rock.

The scale should be just small enough for the mapped area to include a representative variety of bands and any cyclic pattern in the fine banding, while otherwise showing as much detail as possible. This might be, say, 10 m× 10 m, at 1:10, 1:5 or 1:20, to suit the situation. Individual features within the area can be sketched at 1:1 or magnified. These localities should be marked on the over-all field map and on any formation log. The detailed map, in turn, can show locations of individual samples taken, of small inclusions, veins and examples of rocks of which the mineral textures and compositions are noted (as dealt with in later chapters). Large scale maps are particularly useful for bringing out slight angular discordances of rock bands, not visible from any single viewpoint, and for showing variations in the orientation and degree of development of directional fabrics in and around different compositional bands, veinlets, etc. (Section 4.8.3 of 'Basic Geological Mapping': see reference in Table 1.2.)

A *local detailed log* may be constructed, to give a graphical representation of variations in, for example, mineral content across the gridded area. Such a log is not a substitute for a detailed map, because it does not bring out variations along the banding, either slight band discordances, geometrical details of thickness variation, or accompanying changes in fabrics. The time taken to map a small area is limited and relationships intermediate in scale between that of mineral grains and that representable on the map can be spanned by enlarged detailed sketches. For these reasons, detailed maps of example localities should be constructed to show fine banding, whether or not a log is also constructed at this scale.

4.3 Three dimensions

Rocks are three-dimensional, and banding consists of rock sheets, even if they appear as bands on surfaces. However, there may be variation in such things as boudinage and veining with direction along the sheet. Wherever possible, banding should be examined on surfaces having a variety of orientations and any variations in properties with orientation should be reported.

5

Minerals, rock-types, compositions and grades

The matters considered in this chapter provide the basis for every description of metamorphic rocks. This is of such importance that complete sections of the chapter have been devoted to visible minerals (5.1), to fine-grained materials (5.2) and to the naming of rock-types (5.3). These are followed by a briefer statement (Section 5.4) of the general plan which should be followed when describing a rock-type. The final section (5.5) considers the deductions about compositional category and metamorphic grade which may be justified. Tables of detailed information relevant to these sections are given at the end of the book (Sections 10.1–10.3) for easy reference in the field.

5.1 Minerals

Wherever minerals are visible, they provide the basis for both the rock-type names, used for immediate field notes, and the full rock-type descriptions. It is from these descriptions that most statements of origin and grade can be made. Only if minerals are not identifiable will a description use physical properties, such as colour, as the basis for rock-type recognition.

5.1.1 Recording assemblages, proportions and properties

Field notes should be made in such a way that deductions can be produced as easily as possible. Deductions about *grade of metamorphism* may be based on the following.

1 The *existence* of a particular diagnostic mineral.
2 The *coexistence* of several minerals, making up either part or the whole of that group of minerals (*assemblage*) which constitutes the rock.
3 The *absence of minerals or of combinations* of minerals which might have been expected.

Sections 5.5 and 10.1 consider specific grade indicators. It should be noted just how important it is to know *which minerals occur in contact with each other and which minerals do not* (even though occurring separately).

For determining metamorphic grade, the proportions of coexisting minerals are unimportant. However, a fairly close approximation to the rock's *over-all chemical composition* may be made by summing the compositions of constituent minerals, weighted in their correct proportions. In many rocks, composition is the main clue to the nature of the rock before metamorphism, and this can be geologically more important than the metamorphic conditions. Therefore, *mineral proportions must be recorded*.

Clearly, the recording of minerals, their mutual and mutually exclusive relationships and their proportions is a time-consuming and potentially tedious task. However, in addition to justifying deductions about grades and pre-metamorphic rock-types (Sections 5.5 & 10.1), this information may contribute to the syntheses of the rock-mass description, such as the relationships between formations (Section 3.2).

Statements are needed in some cases about *properties of particular minerals*, as well as about the relationships between minerals. This is true if any mineral's properties are unusual, if they cast doubt on the correctness of its identification, or if the mineral is one which often varies in its properties. Things such as quartz with a purple hue, or the colour of a clinopyroxene, should be mentioned. To say that quartz looked glassy, or that calcite effervesced in hydrochloric acid, or that a mica could split into cleavage flakes is a waste of time. As a rule:

1 If a mineral is *recognized*, record it and its proportions.
2 If a mineral is *suspected*, or an identification *doubted*, report why.
3 If a mineral is known to belong to a certain *group*, or to have a *compositional range*, record those characters which are variable for the group or range.
4 If a mineral is *unknown*, describe it fully.

5.1.2 Mineral identification

Minerals visible with a handlens should be identified, or otherwise narrowed down to a particular group with similar properties. The list of minerals (Section 10.2) is based on normal characteristics, of which colour and shape ('habit') are not always reliable. There are many potential variations, which a geologist must beware of, but can also turn to his advantage. For example, an amphibole may occur:

1 In short stubby grains in direct replacement of a pyroxene (Figs 5.1 & 7.11).
2 With needle-like or brush-like habit if it grows in a schist after deformation (Figs 5.2 & 6.1(b)).
3 As a host of equigranular grains in an amphibolite which has accommodated continuing ductile strain (Fig. 5.3).

This list is not comprehensive.

Fig. 5.1 Two pale grains of igneous clinopyroxene in a meta-gabbro. Their strong parting gives a mica-like appearance. The rims have been replaced by dull actinolitic amphibole. Field of view: 26 × 42 mm.

Fig. 5.2 Radiating amphibole needles. (Anthophyllite in an ultramafic schist from the Outer Hebrides.) Field of view: 5×7 cm.

Exsolution, twinning, alteration and *weathering* can all be useful, as well as simple colour. The bronze colour of bronzite is well known. The golden-brown and caramel colours of partially weathered biotite can be just as distinctive. Igneous pyroxenes often possess one main parting along exsolution lamellae, which can give them a mica-like appearance (Fig. 5.1), except that they will break (with difficulty) into small angular granules instead of splitting into flexible flakes. The surface of a large weathered carbonate crystal can look furry, due to the etching of traces of sets of intersecting cleavages or twin lamellae. An orange or rusty deposit on or around weathered carbonate grains indicates a ferroan variety. A surface tarnish or stain on exposed rock is often indicative of sulphide–mineral content, with the stain deeper on sulphide-richer portions and around individual sulphide grains. All these examples are

of general characters. Others may result from local weathering or hydrothermal conditions, and it can be worthwhile to draw up a list of such features for a field area.

Fig. 5.3 Equigranular hornblende, interspersed with plagioclase, in a garnet amphibolite. (Outer Hebrides.) Field of view: 23 × 16 mm.

It is rare for each grain of a mineral to be individually identifiable. Most minerals are identified by putting together mentally the features visible in a number of grains. This assumes that the grains are already known to be the same mineral and highlights the extent to which human vision may perceive likenesses and differences despite an inability to identify minerals. This ability should be used consciously to *extrapolate* identifications from one rock to another. For example:

1 A mineral may be *idiomorphic*

41

(having its own shape), and identifiable when surrounded by one set of minerals, and clearly identical in colour and weathering characteristics to shapeless grains elsewhere.

2 Grains may be larger in some patches, such as veins, than elsewhere, and so display features like cleavage more clearly.

3 Idiomorphic vein minerals may be crystallographically continuous overgrowths on smaller shapeless grains at the margin of the host rock.

Whenever identification in a certain group of rocks relies on extrapolation, this should be noted at the time and reported.

Accessory minerals (those in quantities so small as to be neglected in considerations of rock names) should be reported. It is good practice to search for holes on weathered surfaces, where soluble minerals may have been leached out, and to search in veins and coarse-grained patches. If minerals are found, then they are likely to occur in the rock as a whole, and a search is really to discover whether their grain-size is large enough for them to be visible.

Mineral colour can be a problem. The majority of minerals in metamorphic rocks come in shades of green. Specific mineral greens which should be learned if at all possible are: jade, actinolite, olivine, serpentine, chlorite, epidote and prehnite. (I carry some of these with me in the field.) Though some colours are true mineral colours, others are due to inclusions (e.g. whitish sericite or reddish iron oxide or blackish ilmenite in feldspars). At other times, the colour is that of the surrounding material, seen through a transparent grain (e.g. green imparted to quartz-feldspar rocks by interstitial phengite or sericite). The colours quoted in many books are for thin sections. A mineral with colour in thin section will be blackish to the naked eye, whereas a mineral with colour to the naked eye may be colourless in thin section. True mineral colours are imparted by certain elements only. Minerals containing only elements with atomic numbers up to that of calcium, or these elements plus alkali metals or alkali earths, are basically colourless. Minerals having transition elements or heavy metals as essential components (roughly speaking mafic and heavy minerals) are dark. A small amount of transition element substitution into pale minerals gives bright colours (e.g. small amounts of iron in olivines, phengitic micas, tremolitic actinolites, epidotes and prehnites). A duller effect is usually produced by finely disseminated pigments (e.g. ferric oxides) in a pale mineral base. A few elements are associated with a particular colour (e.g. chrome with green). In this context, the ochre colours of ferric oxides and hydroxides are abnormal. Ferric iron is usually associated with green in silicates (e.g. epidote) and green is not an indication of chemically reduced condition.

Basic mineral *symmetry* may be much more clearly apparent from cleavages than from *shapes*. However, grain shapes may be considered in conjunction with other features. For example, a circular-sectioned rod-shaped mineral probably has more than one symmetrically equivalent direction in the plane of the circular section, and so belongs to an optically uniaxial system. However, if it possesses just one cleavage along its length, such symmetry is refuted, and the mineral must have orthorhombic or lower symmetry. Such quick checks can prevent a number of misidentifications.

Cleavage, and particularly the angle between cleavages, may be observable on a roughly broken mineral surface cutting the cleavage orientations. The surface consists of numerous minute steps along the cleavages, which, though they may be invisible, can still all reflect the light in the same directions. Given a directional light source, such as the sun, it may be possible to 'catch the light' off different cleavages and to make a rough measurement of the angle between them. In particular, this should be a standard approach for distinguishing between similar amphiboles and pyroxenes.

The greatest problem with minerals is often seeing them at all. Some minerals of below or above average resistance to solution weathering may show up on weathered surfaces. Others may be distinguishable by colour only when fresh. Some may show on lightly weathered surfaces by etching of cracks around their grain shapes. The roughness of dry rock may scatter light and show up some grains. Others may be more visible on a wet surface, which scatters light less. All these things depend on local conditions. The only rule is to keep trying, with a hammer, handlens, water, hydrochloric acid and a streak plate.

5.2 Fine-grained material

5.2.1 Fine-grained patches, pseudomorphs and mylonite streaks

Portions of fine-grained material may occur within a rock in which mineral grains are otherwise visible. The two general questions to ask in such cases are, 'How did such a fine-grained entity arise?' and 'What is it made of?'. The possibilities are that:

1 Such portions existed as fine-grained entities in the *pre-metamorphic* state.

2 Grain-size has been reduced by chemical processes of metamorphism (*pseudomorphing*, in a broad sense).
3 Grain-size has been reduced by deformation.

Fine-grained chunks, either rounded or angular, having forms like large clasts are usually pre-metamorphic. Generally they were *lithic clasts* in coarse sediments, *bioclasts* or *concretions* (chert, ironstone or calcareous) in fine sediments, *xenoliths* in igneous rocks. Almost all fine-grained rock-types can occur as lithic clasts or xenoliths. When found in metamorphic rocks, these fine materials may be treated and described as metamorphic rocks in their own right (Section 5.2.2), in intimate association with the coarser rock-types in which they lie.

Many fine-grained chunks, the size of large crystals, showing consistency of composition, size, shape and distribution through the rock are usually *pseudomorphs* (see figures in Section 7.3). If they are aggregated, or vary in size, this will be in the same way as the grains of the mineral they pseudomorph (e.g. clusters of plagioclase or pyroxene grains in a gabbro, or strings of phenocryst minerals parallel to the walls of a dyke). The pseudomorphed mineral may be metamorphic or pre-metamorphic. Although many pseudomorphs have distinctive angular shapes, this is not essential. Just as minerals, of metamorphic rocks in particular, are commonly ovoid or irregularly polyhedral, so will be pseudomorphs formed from them. Beware of the capacity of some minerals to form porphyroblasts by including large numbers of much smaller grains of other minerals. These can look like pseudomorphs, because of the finer grains within them, but they are not. The smaller grains inside

43

such porphyroblasts are usually more similar in mineral species, size and shape to the grains of the matrix than is the normal case for grains of a pseudomorph (Figs. 5.4 & 7.10). Late modifications to shapes may have occurred. Reaction tends to round off the corners of angular pseudomorphs. Deformation may do the same, and may also create an array of strained shapes which could not have accommodated the earlier pseudomorphed mineral. Pseudomorphs may also be zoned. Pseudomorphs are considered in more detail in Section 7.3.

Fig. 5.4 Inclusion-filled ovoid porphyroblasts (probably cordierite). Note the similarity of the fine grains showing inside and outside the porphyroblasts. (Origin unknown.) Field of view: 35 × 24 mm.

A *fine-grained matrix to coarser grains or to chunks of coarser rock* will probably be in a rock looking like either a metamorphosed clastic sediment, or a meta-volcanic rock. It may be just that, but it may also be a product of deformation. Distinction can sometimes be difficult. Whether the fine material is pre-metamorphic or pro-

duced by deformation, the coarser parts are likely to have rounded-off edges, and the matrix is likely to be either dark and slaty, or felsitic, and generally nondescript. In either case, the matrix to quartzo-feldspathic material is likely to be slaty or sericitic, whereas the matrix to intermediate or basic igneous material is likely to be more chloritic. Grain-size reduction may be proved by finding either pieces of an original coarse grain, pulled apart parallel to the foliation, or a streak of fine material cutting through a coarse grain. This grain-size reduction may still have been superimposed on a previous rock with a fine matrix. *The distinction between these possibilities is best made by mapping out their distribution, and by a comparison of associated rock-types.* This should show whether their occurrence is within a stratified depositional sequence (clastic, volcanic or volcaniclastic) or in a deformation zone. Rocks of deformation zones, particularly cataclasites and mylonites, are the subject of Section 9.2.

5.2.2 Fine-grained rocks

Rocks which cannot be described in terms of individual minerals on account of too fine a grain-size should be described in terms of their physical and simple chemical properties: *hardness; colour* and *lustre* on fresh and weathered surfaces; *resistance to weathering relative to other specified rock-types;* degree of visible *anisotropy,* and any *cleavage;* whether *homogeneous, patchy,* or *striped* (either parallel or oblique to any bedding or banding); *whether attacked by hydrochloric acid; whether gradational in grain-size* through to coarser rock-types, and if so, which other properties share in the gradation. These matters overlap with fabrics and cleavage as treated in Chapter 6.

Table 5.1 lists some fine-grained rock-types. Note that, apart from serpentine (or 'serpentinite'), these are either fine-grained before metamorphism, or the pre-metamorphic rock is texturally destroyed by deformation. *Fine-grained siliceous rocks* are not clearly distinguishable one from another, and no suitable and generally recognized common term (roughly the metamorphic equivalent of 'felsite') exists. The term 'greenstone' is not always limited in its usage to fine-grained rocks.

5.3 Rock-type names

The field names of metamorphic rock-types are generally of three kinds:

1 Based on *metamorphic minerals*. These should be used whenever metamorphic minerals are visible.
2 A metamorphic term appended to a non-metamorphic rock-type name.

Table 5.1 Some fine-grained rock-types

SLATES (varieties formed from fine-grained sedimentary rock.)
> *Soft slates* or phyllites, consisting almost entirely of sheet-silicate minerals (sericites and chlorites).
> *Siliceous slates*, with a substantial content of hard quartz (and sometimes feldspar) with sheet-silicates.
> *Marly slates*, containing detectable calcite, and sheet-silicates.

ALTERED IGNEOUS ROCKS (which may be *isotropic* or *slaty*)
> *Serpentine* (or *serpentinite*): from an ultramafic rock.
> *Greenstones*; from a fine-grained basic or intermediate rock. The colour is usually that of chlorite or actinolite. Similar *brown* or *grey* rocks are possible at *Very Low Grade* among apparently unmetamorphosed sedimentary rocks, or in a lava pile.
> *Red-and-green rocks* (no standard name applies), having patches rich in carbonate, patches which are hematite-red and patches which are epidote-green; formed by metamorphism of hydrothermally altered volcanic materials.

HORNFELSES AND SIMILAR ROCK-TYPES
> *Hornfelses* of all kinds. Hornfelsing (turning harder, brittler and more isotropic) is not restricted to a particular previous rock-type.
> *Meta-cherts*. Can be hornfels-like; black, hematite-red or bleached; amongst other metasediments which are not hornfelses.
> *Hornfels-like siliceous rocks* (no standard name applies) amongst other metamorphosed, but not hornfelsed, sediments and lavas, etc.; from igneous 'felsites' or fine-grained sandstones (arkoses or quartzites).

FAULT-ROCKS AND MYLONITES
> *Fault gouge* (incohesive), *pseudotachylyte* (glassy), *cataclasite* (isotropic) and *mylonite* (finely banded, sometimes slaty); from all kinds of rock-types.
> In fault or mylonite zones. See Chapter 9.

These are used when *minerals are visible but their grains are primarily pre-metamorphic.*

3 A summary of the main physical properties. These are used for rocks in which *minerals are too fine-grained for individual identification.*

Other names can be used for special rock-types, in particular those of fault-zones and deformation zones (see Table 9.1).

5.3.1 Mineral-based names

Most metamorphic rocks belong to one of the standard types defined in Section 10.3. *Use a name on this list whenever possible,* but add to it (see below). These names have generally accepted meanings in terms of mineral contents, and should not be used in any other way. For example, an amphibolite is a rock which contains plagioclase and hornblende, so 'amphibolite' should not be used for rocks consisting of an amphibole alone. Additional minerals should be added as *prefixes* to the rock-type name-base, e.g. diopside marble, garnet amphibolite. In the case of micaschists and marbles, the name of the mica or carbonate mineral, respectively, can be added as a prefix to make the name more specific, in which case the name is not taken to be that of an additional mineral.

Rocks which do not fall into these common categories can be given names by adding a textural type-name (schist, gneiss, hornfels) after the name of the most common mineral or mineral pair (e.g. hornblende schist, chlorite-actinolite schist). If no obvious textural term applies, the suffix '-ite' or the word 'rock' is added to the name of the most common mineral, and other

mineral names may precede it (e.g. garnetite, diopside hornblendite).

5.3.2 Names based on pre-metamorphic type

The most basic of these names are 'meta-' prefixed to a previous rock-type name (e.g. meta-arkose). Whenever possible *a textural term should be included,* to indicate the nature of the change which has occurred (e.g. *schistose* granule conglomerate, *hornfelsed* micaschist, *gneissic* meta-granite, *flaser* gabbro). The 'meta-' prefix may be dropped, if to do so does not leave an ambiguity. The name 'granite gneiss' is intentionally ambiguous, but is justified only where it is really unknown whether deformation has occurred before or after the solidification of the rock. Otherwise an unambiguous name should be used. In the case of a granitic rock which has definitely undergone deformation as a solid rock, the term 'gneissic meta-granite' is correct.

5.3.3 Names of fine-grained metamorphic rocks

The basis of fine rock-type names is either physical: *hornfels, slate, mylonite,* etc.; or compositional: *marble, quartzite, greenstone, serpentine.* Both terms should be used if possible (Table 5.1). These name-bases should have added to them words denoting additional over-all properties such as colour (e.g. black slate) or minerals visible (e.g. pyritiferous slate). Sometimes over-all properties have compositional implications (e.g. spotted hornfels, red slate), although these cannot be specified in detail.

5.4 Reporting rock-types

A *fine-grained rock-type* should be introduced by its name, and then described as fully as possible. See Section 5.2.2.

A *rock-type in which minerals are visible* should be introduced by (1) *its name*, followed by (2) *a list of constituent minerals* in order of decreasing abundance. Each mineral name may have written alongside it a word such as 'porphyroblastic' or 'idiomorphic'. If mineral proportions are uniform, these should be written next to the mineral names (as percentages). If any mineral is distinctly a remnant from an earlier rock-type, this too should be indicated.

The rest of the rock-type description should work from the general to the particular, starting with (3) *over-all rock properties* (colour, fracture, jointing, schistosity, weathering style, etc.). There should be (4) *sketches of any compositional patchiness* or fine-scale banding, annotated with statements of mineral proportions in the different parts. Then should follow (5) *sketches of grain textures,* annotated with mineral names. All sketches should show scale, and the orientations of any directional features. Two separate strands of rock-type description lead off from the sketches of textures: mineral details (as below); and fabrics, structural context, etc. (as listed in Section 10.4, and treated more fully in Chapter 6).

Either as annotations to sketches, or separately, there must be (6) *statements about which minerals sometimes occur in mutual contact, and which do not.* It is worth checking that mineral proportions, grain-sizes, and mutual contacts have all been recorded fully up to this point in the description. When this is done, there should be (where necessary: see Section 5.1.1) (7) *brief descrip-tions of each mineral*, and then (8) *descriptions of any fine-grained portion* (Section 5.2.1). For this purpose, minerals can be considered to fall into four groups (as in Section 10.2), to be reported in this order:

1 *Common bulk minerals.* Between them, these usually make up most of the rock. They normally belong to the following list: quartz, feldspars, olivines, pyroxenes, amphiboles, micas, carbonates, serpentine minerals, talc, chlorite and epidote.

2 *'Metamorphic minerals'.* Though chemically composed of the same components as bulk minerals, they are generally present in smaller amounts, and are sometimes individually distinctive of particular metamorphic grades or compositional categories. This group includes: garnets, aluminium silicates, chloritoid, staurolite, cordierite, prehnite, zoisite and lawsonite. *In marbles,* talc, amphiboles, pyroxenes and olivines can be considered in this category.

3 *'Accessory minerals'.* These carry minor chemical elements (boron, copper, titanium, etc.) in concentrated form. They include the oxides of iron and titanium, sulphide minerals, sphene, apatite, beryl, tourmaline and chromite. Frequently, these exist in a metamorphic rock at the same location as did accessory minerals (rich in the same elements) in the pre-metamorphic state, and a note should be made of any evidence suggesting that the present grains existed before metamorphism.

4 *Minerals of unknown identity* (if any). These should be described as fully as possible, with sketches of habit, cleavage, alteration to other minerals, etc.

5.5 Compositional category and metamorphic grade

Indications of previous rock-type may be shown by sedimentary or igneous structures, and by grains of minerals which have not been destroyed by metamorphism. This information should be checked against, and incorporated with, information available from present mineral constitutions, to give an idea of the nature and origin of the rock unit. This section provides a framework for inferring classes of pre-metamorphic compositional rock-type group and of metamorphic grade from the present minerals.

5.5.1 Grade of metamorphism

Changes in the character of a rock as it is traced inwards, through a metamorphic aureole or a metamorphic region, towards the igneous contact or central zone, are called changes in grade. Of these many changes, it is the changes in the assemblages of minerals which can be allocated to defined ranges of pressures and temperatures of metamorphism. In most cases, increases in grade correspond to increases in temperature of metamorphism. This does not mean that *grade* and *temperature* are synonymous. The same transition from one set of minerals to another may occur in different rocks at different temperatures. However, we do not see temperatures. Of those things we can see, the transition in the minerals is the best information we have for saying that different rocks have something in common, i.e. they have the 'same grade'. In other words, *pressure and temperature are measures of conditions which cause metamorphism. Grade is a measure of the effect* (metamorphism).

Mineral reactions differ in their response to *pressure*. Generally, at higher pressure, a mineral reaction will occur at higher temperature, but the extent of this is different for each reaction. Therefore, the transition from one grade division to another, as defined (or at least measured) by different rocks, cannot coincide exactly at all pressures.

In describing a rock-mass, transitions from one set of minerals to another should be noted, and if possible mapped out. The question always arises, 'Does this change in minerals represent a change in bulk chemical compositions, or in grade of metamorphism i.e. an "isograd"?'. Therefore, deductions about composition and about grade must be tackled together. Both must be tackled at two levels. Variations of rock-type *within the rock-mass* can be described and explained either in pre-metamorphic compositional terms, or as changes in metamorphic grade on a local scale. These relationships among the rocks of the rock-mass are vital to the geological synthesis of the area. The other level is that of *external reference*, comparing and categorizing the rocks in terms of an external, generally understandable scheme. Section 10.1 provides such a scheme. It defines compositional classes for common metamorphic rocks, and gives information for classifying their grades. Determinations of pressures and temperatures of metamorphism are beyond the scope of non-specialist description, but some initial indication is provided by classifying metamorphic associations into *Low Grade (LG)*, *Medium Grade (MG)*, etc. (The approximate pressure and temperature fields are shown in Fig. 5.5.) These are broad divisions. If some rocks are partially as under *Low Grade*, and partially as under *Medium Grade*, there need be no hesitation in saying they are transitional between the two. The compositional categories also have imprecise boundaries. For example, the transitional composition between micaschist with and without K-feldspar is slightly different at different grades. It cannot be over-emphasized that this categorization is a simplifying deduction, and is not a substitute for describing those features of a rock which are seen in the field.

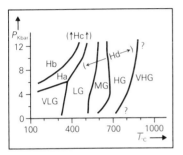

Fig. 5.5 Approximate pressure and temperature fields of the grade categories and high pressure association used in this book. (See Table 10.1.)

5.5.2 Compositional category

Section 10.1 covers all the common metamorphic rocks formed from massive pre-metamorphic parents. It excludes most reaction-zones, metasomatic rocks, hydrothermal deposits and metamorphosed soils and subsoils.

Meta-igneous rocks are divided into ULTRAMAFIC (A), BASIC & INTERMEDIATE (B) and ACID (C). Meta-sediments are divided into QUARTZITES (D), SEMI-PELITE & PELITE (E), CALCAREOUS (F) and META-MARLS & CALC-SILICATES (G). *Pelites proper* are a sub-class (E+). A number of transitional types are possible between the categories used here, but the information should be sufficient to identify

their pre-metamorphic type and their grade. Large clasts, xenoliths, intrusions or cumulate layers of one composition may exist within a host rock of a different category.

The summary (Table 10.1) shows the most common minerals of some rock-types at various grades. These minerals are not necessarily grade indicators. The grade indicators are given in Section 10.1. These may be *individual minerals*, or a *combination* of two or more minerals, which may be only a part of the complete *assemblage*. The word combination is used here when a grouping of minerals indicates grade, but does not constitute the whole rock.

Brief notes have been incorporated into Section 10.1 to minimize cross-referencing with this text. The following notes are on matters requiring more attention.

5.5.3 Basic and intermediate rocks (B)

Igneous plagioclases metamorphose to calc-aluminous portions, and large igneous mafic minerals to mafic portions, with little interaction. Any quartz remains stable, as does biotite except at *VLG* or *VHG*. Therefore:

1 Approximately, the ratios MAFIC : CALC-ALUMINOUS : QUARTZ : BIOTITE correspond to those of the igneous rock.
2 Igneous structures, which are variations in amounts of rock portions, remain visible.
3 Coarse plutonic textures remain as pseudomorphs or augen, etc. (Composition plus structure can closely identify the igneous parent.)
4 Rocks of homogeneously intermixed calc-aluminous and mafic minerals derive *either* from fine-grained igneous rocks, *or* from

49

coarse-grained igneous rocks by deformation, *or* from rapidly re-sedimented igneous material. (These are all likely to be banded in some way.)

5.5.4 Pelitic rocks (E+)

Traditionally these rocks have been zoned by first occurrence of minerals, including biotite, garnet, kyanite or andalusite. However none of these minerals is reliable if rock compositions are highly variable. They may be used to define zones locally if it can be demonstrated *in the field* that they give a consistent zonation. They may even then not allow very close reference to universal grade categories (as in Section 10.1).

Within *VLG*, both grade and compositions produce variability of mineral assemblages, but as several of the minerals are 'sericites', distinctions are not possible in the field.

If carbonate minerals occur as bulk minerals in rocks at *VLG* or *LG* which would otherwise be of this group, such rocks should be classified as 'meta-marls' (G), because they may give rise to quite distinct calc-silicate rock-types at *MG*, *HG*, and *VHG*.

5.5.5 Marbles and calc-silicates (F & G) and fluids

In silicate rocks, prograde metamorphism proceeds by dehydration. If retrograde reactions occur, they re-hydrate higher grade minerals, producing minerals which had previously occurred in the prograde metamorphism.

In marbles, prograde metamorphism proceeds by decarbonation. Retrograde hydration reactions produce hydroxides and hydrous silicates which did not occur prograde. Therefore, some late *Low Grade* minerals are good evidence of earlier higher grade metamorphisms.

Marls contain both silicate and carbonate portions. The silicates dehydrate substantially through *Very Low* and *Low Grade*, while decarbonation reactions occur mainly at *Medium* and *High Grade*. Through *Medium Grade* in particular, carbon dioxide and water dilute each other. This tends to reduce the temperatures at which minerals first appear, and can cause both sides of an expected reaction, such as:

anorthite + calcite + quartz =
   grossular garnet + carbon dioxide

to persist through an interval of metamorphic grade, as measured by other rock-types.

For practical purposes, therefore:

1 Grades indicated by particular minerals are not the same in pure marbles, impure marbles, and in silicate rocks.

2 It is important to record whether a carbonate mineral exists in each calc-silicate rock-type.

3 The most informative calcareous assemblages are those with most minerals.

4 A scheme for determining grade of metamorphism, such as is given in Section 10.1, will not apply to calcareous rocks being metamorphosed a second or subsequent time.

5 Minerals found near contacts (or in cross-cutting veins) of marbles and of silicate rocks may occur at anomalous grades as a result of carbonate and hydrous fluids diluting each other.

6 Generalizations about grade can only be approximate for these rocks.

5.5.6 Basic calc-silicate composition overlap (B & G)

The reactions in calc-silicates which incorporate the calcium and magnesium from carbonates into silicates (such as plagioclase, actinolite, diopside) can turn dolomitic marls into meta-igneous types of mineral assemblage. However, these are compositional flukes within the considerable variability of mixed source clastic/carbonate sediments. Calc-silicate and meta-igneous rock-sequences should be distinguishable by their different modes of compositional variation:

1 Uniform meta-igneous types display a degree of compositional control which is only explicable in terms of magma genesis, not sedimentation. They are meta-igneous.
2 Rocks which vary only in the proportion of calc-aluminous to mafic material, and may contain some quartz and biotite, show a degree of control in their mode of variation which can be explained in terms of igneous mineral segregation, not sedimentation. They are meta-igneous.

(The features mentioned under (1) and (2) may be retained in rapidly re-sedimented igneous material or in volcaniclastic rocks, in which sedimentary structures and bedding may be visible.)

3 Rocks which range in type from highly calcic (rich in carbonate or diopside), through meta-igneous types to meta-clastic rocks of category E (often rich in quartz and biotite) show the kind of variation of meta-marls. They are meta-sedimentary calc-silicates (G).

4 More restricted calc-silicate compositions can develop by metasomatism or in reaction-zones (Chapter 8).

5.5.7 Loss of white mica + quartz: 'High Grade'

In most cases, the breakdown (prograde) of *white mica + quartz*, either by partial melting (C) or by reaction to *K-feldspar + Al-silicate* (D & E), provides a useful definition of the transition between *Medium* and *High Grade*. However, in calc-silicates (G), these minerals breakdown by reaction with calcite *within Medium Grade*:

muscovite + calcite + quartz = K-feldspar + anorthite + CO_2 + H_2O

An aluminous mineral which is calcium-free must coexist with K-feldspar to be diagnostic of High or Very High Grade.

5.5.8 Very High Grade

This is not a coherent grade category in the same way as the others. It is a handy class into which to put a number of rock-types which are dehydrated beyond the extent of usual *High Grade* rocks. Such rocks may belong to a number of unusual associations which are not distinguished in this book. In several of them, normal *High Grade* rocks coexist. It is useful to draw attention to their association by calling it *VHG* if any of the constituent rock-types are as given for *Very High Grade* in Section 10.1.

6

Textures, fabrics, cleavage and schistosity

6.1 General

6.1.1 Terminology

The words *fabric, structure* and *texture* relate to geometric patterns produced by numbers of mineral grains in the rock, at all scales from that of small grain-aggregates to bulk properties. There are several contradictory conventions concerning the usage of these terms for metamorphic rocks. Their meanings in this book are as follows. *Structure* refers to disposition, in three dimensions, of compositionally identifiable portions of rock which are of larger size than individual grains (e.g. bands, spherulites, boudins, beds). *Texture* refers to shapes produced by grain outlines, taking several adjacent grains together (granulose, ophitic, etc.). Whether this is visible in the field will depend on grain-size. *Fabric* refers to over-all directionality imparted to a whole rock by preferred orientation of elements within it (i.e., by the smallest scale structure, the texture, or both). *Microstructure* is not used in the field, being reserved for microscopic and sub-microscopic features.

The word *structure* suffers badly from over-use in geology. Just as people talk of a building having over-all structure, and of its elements, like arches, being 'structures', so too in geology, people talk of the over-all structure of a rock-mass, and of its elements, such as folds, as 'structures'.

Specific elements are classified as sedimentary or tectonic in a way which involves gross simplification and abstraction. A folded bed is a real element of the structure of a rock-mass. In geology, the bed is considered sedimentary (a sedimentary structure in the broadest sense) and the fold is considered a tectonic structure. Then there is the problem of scale, with 'structure' referring to anything from ripple cross-lamination to the lithosphere. In this particular chapter, 'structure' refers to 'over-all structure' at a scale of a hand-specimen, or to 'tectonic structure' (or tectonic structures) at any larger scales.

6.1.2 Rocks without a metamorphic directional fabric

In these rocks, *structure is usually pre-metamorphic,* though a *new grain texture* may be visible. Any directionality is an igneous, sedimentary or compaction feature, and its description can follow naturally from the description of composition. The extent to which a metamorphic textural history may be discerned will be very limited, as will the scope for textural correlations between different rocks.

In the field, as these features are all essentially geometric, their description should be in the form of sketches. Any words needed, such as mineral names, can normally be annotations of a sketch. A scale bar should always be included.

6.1.3   Deformed metamorphic rocks

The organization of material in this chapter has been chosen to suit deformed metamorphic rocks. That is because of the additional geological potential of correlating mineral-chemical aspects of metamorphism with tectonic structures, by way of grain textures. This enables a common (metamorphic and structural) *geological history* to be established. It permits *correlation* of stages of mineral metamorphism from one rock to another by way of common relationships of textures and fabric orientations to tectonic structures, and so it can establish that a geological history is regionally consistent. It may suggest geodynamically important pressure and temperature limits to regional deformation episodes, and can give estimates of gross burial and uplift rates. The ability to specify which minerals relate to which tectonic features can demonstrate the feasibility of drawing conclusions of general significance from further study (of things as diverse as iostope geothermometry and strain-rate determinations).

In the field, if the procedure of this book has been adopted, there will already be a record of general location, fabric and structure at map-scale, any banding, and the mineral content of particular rocks. Textures, etc. can then be recorded as in the *checklist for textures and fabrics* (Section 10.4).

In any compositionally variable formation, it will be clear that there is neither time nor purpose in recording very similar textures, related to the same over-all fabric, in one rock after another. It may be that a particular rock has unusual composition or texture, and that it is worth describing, regardless of its location. Otherwise, rather than choosing examples of textural types from unrelated locations scattered across a formation's outcrop, it is better to choose an *example locality* in which several types occur. This can then be grid-mapped to show how variations in orientation or intensity of fabrics are related to banding, smaller-scale structures, etc. (Fig. 4.9 of *'Basic Geological Mapping'*: see reference in Table 1.2). Accurately located examples can be chosen from within the mapped area for description of textural and other details. Statements are in any case necessary about the extent of outcrop over which such recorded features are representative.

6.2   Textures

All textural matters are fundamentally geometrical, and can be recorded best in sketches. Terms like 'decussate', 'bow-tie', and even 'augen' can be used for such a variety of types as to be almost worthless unless accompanied by sketches. A properly annotated sketch will show the minerals adjacent to those for which such a term might apply, and the relative grain-sizes, habits of other minerals, degrees of mineral association, patchiness, directionality, etc. This is more than could be put into words, however long and thorough a verbal description. (See Section 5.4 for overlap with recording minerals.)

In the field, as a rule:

1 Always sketch.
2 Always annotate with names of minerals.
3 Always include a scale bar.
4 Always record the attitude of the

53

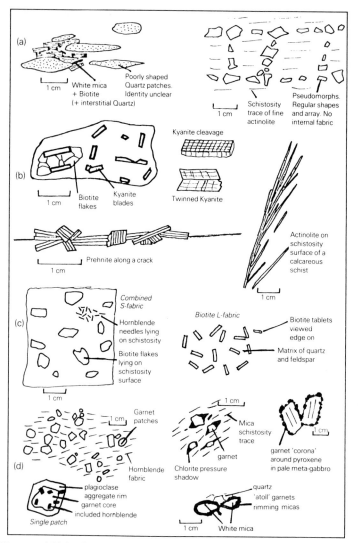

Fig. 6.1 Field sketches of textures. (a) Shape, size and distribution of some compositional patches. (b) Some mineral habits: kyanite, prehnite, actinolite. (c) Some mineral fabrics. (d) Some mineral associations and time relationships. (Garnet examples.)

54

sketched surface, and the pitch of features on it.

Figures 6.1–6.3 show examples of types of textures and some time relationships.

6.3 Fabric, cleavage and schistosity types

6.3.1 Mineral fabrics

Mineral fabrics exist to the extent that there is a preferred orientation of the

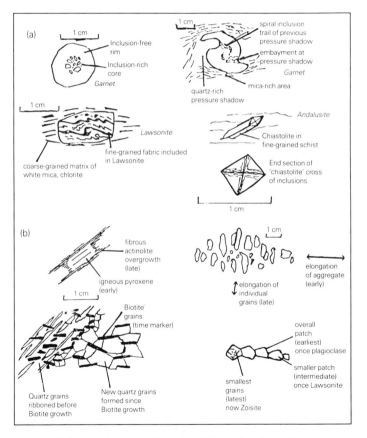

Fig. 6.2 Field sketches of textures. (a) Inclusions. (b) Some time relationships.

mineral grains (Fig. 6.4). The preferred orientation may exist in different degrees for different constituent minerals. In theory there is a distinction between 'grain shape fabrics' (of grains having alignment of external shapes regardless of internal crystal axes) and 'grain orientation fabrics' (of grains having alignment of axes regardless of external shapes). In practice, such a distinction is too simplistic. Many schists, for example, have a visible alignment of mica crystal directions and of both mica and quartz grain shapes. This should be called a 'mineral fabric' and recorded as a sketch, without worrying further about terminology.

Fig. 6.4 Biotite L-fabric (see Fig. 6.6), viewed down the lineation. Most grains are seen edge on, with little preferred orientation in this plane.

6.3.2 Shape fabrics

The other general type of fabric consisting of discrete simple elements is 'aggregate shape fabric', or quite simply 'shape fabric', in which it is compositional patches, of many mineral grains, which have a preferred orientation of their shapes (Fig. 6.5).

6.3.3 Fabric symmetry

Mineral and shape fabrics, because they consist of discrete elements, can possess a so-called 'symmetry'. Any mineral or shape fabric should be classified as 'L', 'S' or 'LS' on the basis of this symmetry (see Fig. 6.6). The same terminology may be used in a less

Fig. 6.3 Dark actinolite blades, with weak fabric, overgrowing quartz ribbons (medium greys) and compositional banding (running top right to bottom left). (Inner Hebrides.)

rigorous way to say whether other types of fabric are mainly linear, planar or mixed in either their resulting fissility or their visual effect.

Fig. 6.5 Shape fabric in deformed and metamorphosed calcareous conglomerate. The apparent fabric trend (the lengths of sections through deformed clasts) pitches less steeply on this surface than the trace of the compositional banding (bedding). Field of view: 1.5×2.5 m.

6.3.4 Fabric origin and grade

Mineral and shape fabrics can have diverse origins. Evidence should be sought in answer to two separate questions.

1 When were the material elements which constitute the fabric formed?
2 Were they formed with, or have they been later endowed with, their preferred orientation?

In meta-igneous rocks, parallelism of grains or aggregates may result from magma flow, gravity settling, compaction, magmatic crystal growth, deformation associated with intrusion, or later tectonic deformation. (This list may not be comprehensive.) In meta-sediments, parallelism may be produced by sedimentation, diagenesis and compaction, metamorphic mineral growth mimetic on a sedimentary fabric, deformation acting during metamorphic mineral growth, or deformation acting upon pre-existing metamorphic minerals. *Answers to questions of fabric origin should be sought from independent textural evidence, before considering relationships to tectonic structures.* Where it is metamorphic minerals which constitute a fabric, they indicate the metamorphic grade of the rock at the time the fabric developed. This may have been later than the time at which that metamorphic grade was first reached.

6.3.5 Lamination, crenulation and striping fabrics

Some rocks possess a very strong planar directionality in the form of continuous or semi-continuous compositional lamellae. In some the lamellae result from finite strains so great as to have turned nearly equidimensional shapes into fine lamellae. In others, directionality results from material segregating directly into compositional stripes (Section 2.2.1). Some directionality is simply thinned-down highly strained pre-metamorphic banding. It is frequently due to more than one of these processes, or to their interaction.

Some geologists say these planar features should not be referred to as fabric. That would be fine if we knew

57

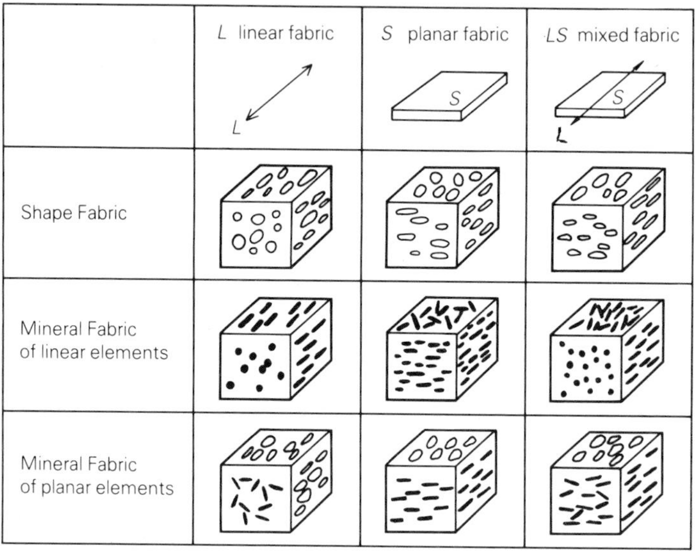

	L linear fabric	S planar fabric	LS mixed fabric
Shape Fabric			
Mineral Fabric of linear elements			
Mineral Fabric of planar elements			

Fig. 6.6 Simple fabric symmetry types.

Fig. 6.7 Folded metamorphic lamination (lamination fabric).

what we were dealing with, but the origins of compositional lamellar fabrics are often uncertain or confusingly mixed. 'Fabric' is allowable, on the basis that (1) a common term is needed, (2) no other term is in general use, and (3) it is bad science to enforce a terminological distinction between features indistinguishable in practice.

Three compositional lamellar fabrics which can be recognized in the field are called here 'lamination fabric', 'crenulation fabric' and 'pressure solution striping'. *Lamination fabric* (or 'metamorphic lamination', or just 'lamination') consists of a continuous pack of parallel compositional lamellae of indefinite origin, usually parallel to banding at larger scales (Fig. 6.7). Such fabrics often contain aligned sheet-silicates, giving a very strong lamination-parallel schistosity. *Crenulations* are sub-parallel thin deformation zones (Section 6.4) obliquely cutting a previous fabric (Figs. 2.1 & 6.8). (It is believed by some, wrongly, that crenulations are universally produced by buckling instability of the previous fabric, and accommodated by material segregation, so making microfold limbs into 'pressure solution stripes'. Where geometry is observed, the preferred name should be the geometrical one of 'crenulation'. Any evidence of segregation should be noted without it contributing to the name.) The term *pressure solution stripe* should be reserved for sharply bounded (commonly anastomosing) compositional stripes which are not subordinate to crenulations (Fig. 2.2). Some stripes occur in rocks which have no earlier fabric to crenulate. Others are more spatially regular than crenulations in the same rock, or correspond to, say, every fifth crenulation only.

In the field, such fabrics should be *named,* and at chosen localities they should be *sketched* to show whatever detail is visible, taking care to include a scale bar and to record orientations. Any early fabric, perhaps crenulated or cut by later pressure solution stripes, should be characterized in as much detail as the late fabric. The *sense of displacement* across crenulations, the *sense of rotation* from early to late fabrics, and their *intersection directions* should be under constant scrutiny. Changes of sense, or of intersection direction, should be located on maps and sections.

Fig. 6.8 A crenulated cleavage surface. The crenulations in this sample are kinks (Fig. 6.16) rather than microfolds (Fig. 2.1).

6.3.6 Slate fabrics

Recognition of the fabric types so far mentioned depends on being able to see the particular elements (grains, crenulations, etc.) that constitute them. In slates all the same types are possible, but they cannot be distinguished without microscopic techniques.

In the field, a note should be made of any compositional streakiness, and the nature of any directionality visible on cleavage surfaces. There might be a *mineral elongation*, a *stretching direction*, the *bedding trace*, and one or more later *crenulations* (Fig. 6.8). Usually, the first two of these will not be separately distinguishable, and the only true evidence of fabric may be the *cleavage*.

6.3.7 Cleavages and schistosities

Slates, phyllites and schists have a directional *fissility* (ability to split easily in a particular orientation) because of an alignment (a fabric) of mineral grains with good cleavage. This allows cracks to develop easily through the rock by passing directly through mineral grains along their cleavages. Sheet-silicates (micas, sericites, talc, chlorites or serpentine minerals) or amphiboles are usually responsible. To present almost continuous paths for the rock to split along, these minerals must either make up the major part of the rock's composition, or be concentrated and aligned along laminations, crenulations or pressure solution stripes (though in fine rocks this may not be visible).

The terms 'slate', 'phyllite' and 'schist' are supposed to be reserved for rocks having a 'metamorphic' fabric. This in practice means that 'slate' may be used for all fine-grained rocks with any sign of fissility which is not due to an unambiguous sedimentary fabric. The terms 'cleavage' and 'slate' are used if the grains responsible for fissility are too small to be individually visible. If there are separate planes down which

such a rock splits, there is a *spaced cleavage* (Figs. 2.2, 6.9, 6.10 & 6.11), of which *crenulation cleavage* and *pressure solution cleavage* are two identifiable types. If adjacent planes of potential splitting are so close that their character and spacing are too small to be visible with a hand lens, there is said to be a *penetrative* or *slaty cleavage*. If a fine-grained rock has fissility in two or more directions, the rock may split easily into long thin pieces (pencils) and be said to possess *pencil cleavage*. If it breaks into flat pieces, it possesses *scaly cleavage*. (See Figs. 6.12 & 6.13.) In these cases, the orientations of the various cleavages should be recorded (if they are at all regular) together with any evidence (orientation, or crenulation of one by another) of relative age. A fine-grained rock with only poorly developed fissility will not split into thin sheets, but will still fracture slightly

Fig. 6.9 Spaced cleavage in quartzite. The spacing of cleavage is widest in the thickest sedimentary laminae.

more easily in one direction than another. This fracture with slightly preferred orientation is sometimes called *fracture cleavage*.

Rocks having grains producing fissility large enough to be visible are *schists,* and their fissility is called *schistosity.* Different schistosity types, equivalent to the different cleavage types, are not normally named, but a note (normally by way of a sketch) should be made of the type of fabric to which the schistosity corresponds.

Fig. 6.10 Close jointing or spaced cleavage (running top left to bottom right) which might easily be mistaken for bedding in calcareous flysch. Actual bedding defines a syncline to which the spaced cleavage is axial–planar. Width of view 1 m.

Phyllites have a grain-size approximately on the borderline between 'slate' and 'schist'. However,

their characteristic flakiness and sheen only develop in rocks of particular, generally highly micaceous, composition. 'Phyllite' is therefore better used as a rock-type name, and not, in the manner of 'slate' and 'schist', as a general grain-size category for fissile rocks of all compositions.

Fig. 6.11 Axial–planar spaced cleavage, roughly corresponding to crenulations of the folded earlier fabric in a silty low grade marble.

6.4 Deformation fabrics traversing a band

The variation in orientation of a fabric in a systematic manner, as it traverses a band of rock, is known in a general way as *refraction*. The band may be a *compositional* entity (a bed, for example) or it may be a *deformation band* or *zone* which is not compositionally defined. Deformation bands are of *shear-zone* or *kink-band* type.

Fig. 6.12 Pencil cleavage. The pencils result from almost perpendicular intersection of a cleavage with some previous fabric (in this case an earlier tectonic cleavage).

Fig. 6.13 Somewhat bladed pencil cleavage. As Fig. 6.12, except that here the intersection of the two fabrics is not perpendicular.

6.4.1 Refraction

The strains suffered by bands of different composition when a banded

rock deforms will differ, according to their competence. This gives rise to variation in the orientation at which fabric develops. Planar fabrics cut the most competent bands nearest to their perpendicular. If the fabric is responsible for a cleavage, this produce *cleavage refraction*. This can be useful in showing *grading, banding* generally and the *way up* of deformed, fine-grained, graded beds (Figs. 6.14 & 6.15). The term *refraction* is often restricted to this usage.

Fig. 6.14 Cleavage refraction through a graded bed in calcareous flysch. The cleavage, visibly axial planar in the lower half of the picture, refracts abruptly at the bed base, at first steeply then gradually flattening back through the bed, showing 'right-way-up' Bed thickness: 1.5 m.

6.4.2 Shear-zones

Deformation zones of shear-zone type are not essentially demarcated by compositions, nor are they limited in their

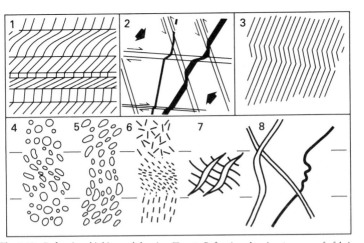

Fig. 6.15 Refraction, kinking and shearing. Top: 1. *Refraction*, showing the traces of a fabric through horizontal bands, one of which is graded. 2. A *conjugate* set of *shear-zones* off-setting dark bands. The arrows indicate displacement sense across individual shear-zones, and the common principal direction of shortening. 3. Traces of fabric through a *kink-band*. Bottom: A sinistral shear-zone passing through unaffected rock, showing (left to right): 4. A new-shape fabric. 5. A re-worked shape fabric. 6. A new mineral fabric. 7. A spaced cleavage (e.g. pressure solution stripes) cut by sigmoidal extensional veins. 8. Passive folding of pale veins and active folding of a dark vein.

geometry by a pre-existing fabric. Several fabric relations may exist in shear-zones (Fig. 6.15). A pre-existing shape fabric will be reoriented according to the finite strain. A pre-existing lamination will be reoriented as a material entity. A new mineral, pressure solution or slaty fabric may develop. Shear-zones exist at all sizes from the sub-microscopic to many kilometres wide. The strain may not die out outside the zone, but simply reduce to a low value.

6.4.3 Kink-bands

A kink-band has some similarity to a shear-zone, except that a previous fabric plays an active role in its development. This is shown by the sharp angular change in fabric orien-

tation at the edge of the band, and the roughly equal angle between this edge and the fabric inside and outside the band (Figs. 6.15 & 6.16).

Fig. 6.16 Kink bands in a micaschist.

63

6.4.4 Significance of shear-zones

Any cases of the fabric relationships mentioned above (or examples of them where numerous) should be sketched, including additional features such as veins (Section 7.4).

In its metamorphic aspects, a shear-zone is the most significant kind of deformation band. The zone represents a band of rock within which deformation has forced continual re-establishment of the mineral constitution. This may have occurred while less intense deformation affected the neighbouring rock, or it may have taken place at a later stage than any changes in the rock which surrounds it. If the rocks within and outside the zone differ in mineral content, they represent different metamorphic conditions which have been passed through during the metamorphic history of the rock-mass. Such a difference in rock-types should be fully described. Shear-zones of low or medium grade assemblages in earlier higher grade or igneous rocks are not uncommon. Shear-zones may be correlated as belonging to the same generation if they share common grade, approximate orientation and sense of displacement (or constitute a conjugate set, Fig. 6.15). Cross-cutting relationships with other structures can be used to place them and their grade into a structural geological history.

6.5 Deformation fabrics and folds

6.5.1 General situation

Consistent fold patterns provide a basis for structural correlation. Any clear relationship among metamorphic minerals, their resulting fabrics and folds, can therefore be extremely important. Synchrony of development of a fabric and of *active* folding is supported by an axial-planar or neutral point relationship. A difference in the time of development is established where a fabric is folded, or where a fabric cuts across and disregards (save for refraction) a previous fold (Fig. 6.17). However, it is quite possible that no consistent pattern of folds or fabrics exists, or that folds are purely passive features. *Passive folds* occur locally in shear-zones (Fig. 6.15), and more generally where masses of heterogeneously deformed rocks lack a consistent orientation of any anisotropy in their physical properties. In general, the higher the metamorphic grade, and the more passive the rock units, the less likely there is to be simple and consistent folding. *In the field*, it is necessary to *show whether a consistent relationship exists or not* between fabric and structures.

6.5.2 Axial-planar relationships

The relationship known as 'axial-planar' (Fig. 6.17, also Figs. 6.10, 6.11 & 6.14) is the most common to characterize fabric and folds thought to be produced during the same deformation. The term has been criticized, because exact parallelism between fabric and a fold's axial plane is almost unknown. In particular, the fabric plane rarely contains the fold hinge. Fabrics also fan through bands of different competence. The name 'axial-planar' may therefore be inaccurate, but to say that the relationship is badly named does not detract from its existence. A planar fabric which develops synchronously with folding will lie closer in its orientation to the fold's axial plane than will any material entity being folded. Therefore, the fabric cuts banding on fold limbs with the same sense of asymmetry as it would if truly axial-planar. *The practical importance of this is that, where it can be demonstrated to exist:*

1. *It justifies correlating the episodes of formation of fabric and of folding.*
2. *Banding/fabric sense can be used to extrapolate from smaller to larger structures, in the same way as parasitic folds.*

It is with this significance that the term 'axial-planar' should be used.

6.5.3   Neutral points

Bending of a competent layer can cause it to shorten on the inside of a fold, but to lengthen locally on the outside of a fold-hinge, even though the wider effect is over-all shortening of the layer. Only between the 'neutral point' and the 'neutral surface' within the layer does a planar fabric lie sub-parallel to the layer instead of the axial plane (Fig. 6.17). Cleavages and mineral fabrics having these relationships are particularly good evidence of synchronous development of fold and fabric, and of the fact that such fabrics lie approximately in the local plane of flattening. *In the field*, it is important to *identify such relationships*, *in order*:

1. *To show that the developments of fold and fabric have been synchronous, and so allow correlation.*
2. *To distinguish them from those of similar geometry where an earlier fabric has been folded around the fold-hinge.* (Fig. 6.17).

6.5.4   Superimposing folds and fabrics

A planar fabric which developed later than a fold will be superimposed on the fold geometry already established, subject to normal refraction and fanning around areas of contrasting ductility. This might conceivably produce a relationship which is again axial-planar (to the earlier fold) if there is near coincidence of strain orientations of the two deformation episodes. If the potential new fabric orientation lies only a few degrees from an existing one, it is normal for the earlier fabric to be redeveloped instead of a new fabric developing. Thus, two deformation episodes can result in only one fabric, axial-planar to one generation of folding, so that the episodes cannot be discerned on either fabric or fold evidence. Even more unlikely are some cases of a distinguishable new fabric, axial-planar to earlier folds. These are *geometrically* discernible in those situations shown in Fig. 6.17. *Metamorphic* distinction is possible where minerals of the two fabric generations coincide in orientation, but differ in metamorphic grade. Apart from these unusual coincidences, fabrics developed later than folds will show cross-cutting relationships.

6.5.5   Only correlate like with like

When using relationships between fabrics and folds, it is important only to correlate like with like. In terms of fabrics, this means that *mineral fabrics* belonging to *particular deformation episodes* should not be correlated with *shape fabrics* which record the finite strain accumulated through *all episodes*. Fabrics should not be correlated on orientation if they are produced by minerals of quite different *metamorphic grade*.

In terms of *folds*, it is important to realize that fold orientations will only be constant over those masses of rock which presented *the same pre-fold anisotropy orientations to the same stress orientations*. In particular, some

65

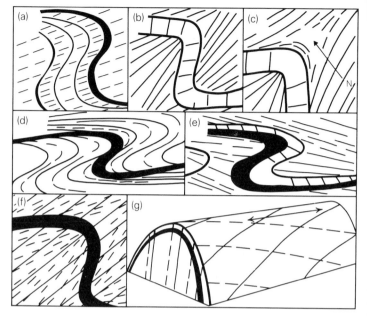

Fig. 6.17 Fabric/fold relationships on profile sections. Thin lines represent fabric traces. (a) Axial–planar. (b) Axial–planar with fanning. (c) Neutral point (N). (d) Fabric earlier than folding. (e) Fabric later than folding showing refraction (one possible configuration). (f) Superimposed axial–planar fabrics, distinguishable by degrees of fanning. Extremely rare. (g) Superimposed axial–planar fabrics, distinguishable by divergence on the banding surface, but not in profile. Rare. Arrow indicates hinge direction.

folds may develop by buckling veins or intrusions, others by buckling banding and some by buckling an earlier fabric. These may all share an approximately common axial-planar orientation, with an axial-planar fabric, but they may differ in fold asymmetry, sense to banding and hinge direction, so that they cannot all be used together to establish larger structures. Veins, in particular, are notorious for producing folds oriented differently from those of banding or fabric. The features which appear to actively create the folding, and those which appear to be passively folded by it, should be noted.

7

Scattered entities: pods, boudins, augen, pseudomorphs, veins and pegmatites

This chapter deals with objects which may be found contained within metamorphic rocks. Its first two sections concern lens or pod-like objects in rocks which have been deformed — Section 7.1 with pods of different rock-type, Section 7.2 with individual large mineral grains, augen, etc. Section 7.3 is about pseudomorphs — patches having the shape of individual grains of one mineral while consisting of one or several other minerals. Section 7.4 deals with 'veins' in the broadest sense, and purposely makes no distinction in treatment between those (such as pegmatites) of known igneous origin and those of unknown or of non-igneous origins.

7.1 Boudins and shear-pods

Within a deformed and metamorphosed formation, there may be blocks or slabs of rock which have suffered much less strain than their surroundings (Fig. 7.1). If they make up discontinuous but recognizable layers or intrusions, broken up by boudinage or shear-zones, they can be described as part of their host formation: care must still be taken with recording of edges and reactions, as noted below. If they are isolated or scattered masses, they deserve individual attention.

1 Do *rock compositions* indicate that the material was of very different origin from the rest of the formation or unit in its pre-deformed state, such as an igneous intrusion in meta-sediments? (Compare compositional categories, Section 10.1.)

2 Do the *location and shape* suggest that the less deformed mass is of initially more competent composition than its surroundings; or that it is different because localized tectonic structures have produced areas of lower and higher strain from a similar set of materials? A *boudin* or *pulled-apart block* (Fig. 7.2) is typical of more competent materials. A *shear-pod* (a volume bounded by zones of higher shear strain) need not have been any more competent than adjacent rocks of the same composition, but is commonly bounded partly by less competent compositions and partly by shear-zones in the same compositions (Fig. 7.3). A *fold hinge-zone* commonly becomes an area of lower strain in more strongly strained rocks of similar composition on the fold limbs.

3 Do the *same minerals and similar texture* occur in the less deformed

Fig. 7.1 An undeformed pod. The pod, 1.5 metres long, is wrapped around by the foliation of the surrounding deformed rocks.

Fig. 7.2 Pulled-apart blocks of a dark competent band in a pale, more ductile marble. (Andalucia.)

block and its more deformed matrix (suggesting that the processes creating the mineral constitution have continued to operate throughout areas of different strain), or is the less deformed block a remnant having an early constitution which have been destroyed in the matrix during deformation?

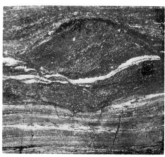

Fig. 7.3 A shear-pod. Two sides, top left and bottom right, are compositionally defined. The banding in the pod passes into sheared contacts on the other two sides. Pod width: 35 cm.

In the field, the first notes about a less deformed block should include the answers to these questions, and a sketch of the shape of the block and the traces of any banding and fabric through the block and its vicinity (Figs. 7.3–7.5). If more than one block of similar type exists, a note should be made about whether there is any consistent relationship for the set, particularly of the 'sense' of angular discordance between the fabric, banding and the boundaries of shear-pods (Fig. 7.3).

Following notes should give evidence of whether the less deformed rocks provide a new and as yet undescribed composition, texture, fabric (Fig. 7.4) or set of small structures. If these show a grade of early metamorphism, or an early tectonic event, they may play a large role in the establishment of the

Fig. 7.4 A pod retaining an early coarse quartz fabric with early orientation. The surrounding more deformed and finer-grained gneisses have a new fabric.

geological history of the rock-mass.

Finally, relationships at the block's edge should be thoroughly described (Figs. 7.5 & 7.6). Some blocks may be surrounded by a reaction-zone (Section 8.2). Bands within the block may pass into the matrix, or be cut off by it. Sometimes, some bands trace into the matrix while others do not (Fig. 7.6). This may reflect competence contrasts, or may result from thinning down of a band bounded by reaction-zones to a point where only the reaction-zone exists in the highly deformed matrix.

7.2 Augen, flaser and large mineral grains

7.2.1 Augen

Augen (meaning 'eyes', plural) are patches of distinct composition, each of which was at some time (in most cases) a single grain, either a crystal or a clast, and now has the appearance of an 'eye' shape on a rock surface. The shape consists of a more equidimensional core, mantled by (or having tails of) a deformed aggregate of mineral grains which, although streaked out, are still part of the compositionally defined patch. The core may be a single crystal, a remnant of the grain which originally constituted the patch. The mantling grains may be either of the same mineral, or of a new mineral with similar chemical composition. In three dimensions, the entire patch may be spindle-shaped, discus-shaped, or something between the two. Because the cores may preserve an early structural state and orientation of an early mineral, mantled by material developed during more recent deformation, augen can sometimes be used in correlations of simple structural and metamorphic rock histories. When this is not possible, they may still provide

69

evidence of the structural relations between earlier and more recent structural states, often in rocks such as gneissic meta-granites, in which little other information exists.

Fig. 7.5 As Fig. 7.4 except that the fabric passes gradationally (as an aggregate shape fabric) from the pod into its more deformed surroundings. A dark mylonitic band of intense strain bounds the right edge. Field of view: 30×50 cm.

7.2.2 Terminology

In igneous rocks, scattered mineral grains which are distinctly larger in size than those of an intervening matrix are called *phenocrysts*. In metamorphic rocks, scattered mineral grains which have *grown* to larger sizes than grains in an intervening matrix are called *porphyroblasts*. Lensoid, spindle-shaped, or discus-shaped compositional patches with pointed ends in

section are known as either 'flaser' (streaks) or 'augen' (eyes). *Augen* are produced by deformation, the sharp ends around a wider centre resulting from strain being localized in their rims. If such a distinct contrast between less strained core and more strained rim is lacking, a thinner *flaser* shape is produced (Figs. 2.4 & 7.8). Augen or flaser may have started as unusually large grains (clasts, phenocrysts or porphyroblasts), or as grains which were the same size as other minerals until the deformation and metamorphism occurred. If *metamorphism* is held to be the main cause of grain-size reduction, remaining large crystals have no special name. If *deformation* is the main cause of grain-size reduction, then remaining

Fig. 7.6 Shear-pod edge. Banding in the fine amphibolite is sheared into the margin, but then truncated by the coarser quartz-bearing amphibolite which also exists as compositional bands (top left) within the pod.

resistant grains are called *porphyroclasts* (Fig. 9.1). Where both have clearly occurred (normally the case), the term porphyroclast is not normally applied.

Pressure shadow compositional differentiation around grains gives them 'elbows' or 'tails', producing a similar overall shape to augen. Unlike augen, pressure shadow materials are usually compositionally unlike their competent equidimensional cores. *Fibrous* or *bearded* growths may occur both within a pressure shadow and at the ends of augen, though exactly parallel *fringes* are restricted to pressure shadows (Fig. 7.9).

7.2.3 Questions

In the field, the items considered here should already have been sketched

Fig. 7.7 Augen of potash feldspar in gneissic meta-granite.

(Chapters 5 and 6) to show shapes, orientations and distribution amongst the other grains of the rock. The task now is:

1 To check that annotations convey all the available information.
2 To list specifically any deductions which can be made about origins and grade, and to state whether there is any local or regional consistency of orientational relationships. (These are matters for correlation with other field evidence.)

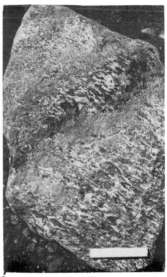

Fig. 7.8 Flaser in meta-gabbro. Dark flaser of actinolite, after igneous pyroxene, and pale flaser, after igneous plagioclase, constitute a deformation fabric which runs top left to bottom right, oblique to igneous grain-size banding.

The following list of questions indicates the information needed.

Are the shapes those of undeformed mineral grains, or the products of

deformation? If the latter, are they:
(1) ductilely *deformed single grains*?
(2) pulled-apart, *broken single grains*?
(3) one of the types *with pointed tails*?
If the last of these, are they *flaser* of one mineral, or do they have a separate resistant core and a deformed rim or tail?

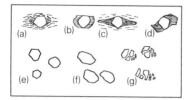

Fig. 7.9 Field sketches of pressure shadows, fringes and grain shapes (all garnet examples). (a) Quartz pressure shadows in micaschist. (b) Pressure fringes, of parallel fibres (amphibole). (c) Bearded pressure shadows or fringes. (d) Rotated pressure fringes. (e) Normal equidimensional grains. (f) Slightly lengthened grains, giving a grain shape fabric. (g) Broken and pulled-apart grains.

In the case of isolated grains, or of resistant cores, is the mineral *metamorphic* or *pre-metamorphic*? If metamorphic, is it of the same *grade* as the rest of the rock?

In the case of deformed patches with resistant cores, are the cores single grains, or the broken pieces of earlier single grains? Are the tails granular or bearded? Are the tails composed of minerals with similar or very different composition to the cores? Are the tails composed of minerals found nearby in small veins? Are the cores and tails of a similar metamorphic grade?

Are *inclusions* of small grains visible? If so, do they show either a preferred orientation of individual granules (a fabric), or concentration in particular lamellae (a lamination)? Is the orientation of either of these:

1 A rational one in terms of the *crystallographic directions* of the host grain?
2 *A fabric, or lamination, which the host grain has overgrown* during its development?

If the latter, is there a change in orientation at the grain edge? Is such a change a sudden kink, or a curve within the grain margin, or both? Is it 'rolled'?

In the case of grains or augen cores which are of minerals having a markedly non-equidimensional shape of any kind, do the shapes have a consistent orientation, or a consistent *sense* of angular variation from a matrix marker, such as fabric or banding? If so, what are the orientations of grains, fabric and banding (as appropriate) and asymmetry sense, and is the pattern consistent for one outcrop, a group of outcrops, or for the formation? Does the sense of asymmetry change across major folds?

7.3 Pseudomorphs

Pseudomorphs demonstrate a time relationship between two mineralogically definable geological episodes, one in which an early mineral was created, the other in which it was pseudomorphed by later minerals. They are therefore important in the establishment of a geological history. Identifying the minerals of the two episodes from their similarities of composition can be very complicated, and this aspect should therefore be tackled last.

7.3.1 Early mineral and the textural evidence

Texture. Textural evidence can often be used to establish the stage in the rock's history at which the pseudomorphed mineral existed. In metamorphosed plutonic rocks, the earliest texture is igneous. Is the pseudomorph an integral part of such a texture? If so, the mineral was igneous (Figs. 7.11 & 8.2).

In coarse metamorphic rocks, the latest texture is that of the present mineral assemblage. Is the pseudomorph an integral part of the present metamorphic texture (Fig. 7.10)? If so, the pseudomorphing is late, and probably a retrograde replacement. In such a case, the early mineral coexisted stably with the other minerals now seen, and its replacement may not be able to coexist with them in stable equilibrium despite lack of obvious reaction.

These are the easy cases, where quite strict limits are placed on possible

pseudomorphed minerals by knowledge of the rest of their assemblage. In other cases, the early stage of the metamorphic history may have to be considered in terms of possible minerals, veining, deformation and grade.

Fig. 7.11 Radiating hornblende crystals, replacing the igneous pyroxenes in a gabbro (between fresh plagioclases with good cleavage and twinning). Field of view: 4×5 cm.

Phenocryst or porphyroblast? A common problem is the identification of large pseudomorphs in an otherwise homogeneous finer-grained meta-igneous rock. If location, size distribution, orientation or deformation state positively prove them to be after porphyroblasts, then the problem does not arise. Examples of this are:

1 Pseudomorphs enlarged or concentrated along small fault-planes or joints (Fig. 7.12).
2 Undeformed pseudomorphs defining a deformation fabric (Fig. 7.13).
3 Pseudomorphs overgrowing a previous fabric.

Fig. 7.10 Pseudomorphs (of zoisite after lawsonite). Note that each is a grain aggregate, as distinct from either a clean single grain or an inclusion-filled single grain (such as Fig. 5.4).

4 Pseudomorphs occurring in hydrothermal veins as well as host rock.

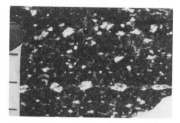

Fig. 7.12 Pseudomorphs, after a mineral
which has grown bigger along a crack, now
healed. Centimetre scale.

Fig. 7.13 Pseudomorphs which are an
integral part of the metamorphic fabric of a
deformed rock. Field of view: 6 × 10 cm.

These all show that a post-igneous process preceded or accompanied the early
mineral, which is therefore metamorphic. If the location, size distribu-

tion and orientation are explicable in
igneous terms, their origin may be
difficult to determine. Igneous minerals are often better developed in (to take
three examples) some cumulate bands
in gabbro, pillow rims in pillow lavas,
or planes running parallel with the wall
of a dyke. These situations are all possible sites for preferential growth of
metamorphic minerals. In cases such as
these, it is normally necessary to rely
on pseudomorph shape, plus knowledge of igneous mineral compositions,
to identify the early mineral (Fig. 7.14).

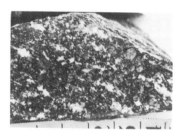

Fig. **7.14** Pseudomorphs of pseudomorphs. Larger white patches (once igneous
plagioclase) broken into short strings of
rhombs (once early metamorphic lawsonite)
each of which is now an aggregate (late
metamorphic zoisite). Centimetre scale.

7.3.2 *Questions about metamorphic history*

In cases where the pseudomorphed
mineral belongs to the rock's metamorphic history, it may be possible to
make definitive statements about its
timing in terms of textures and fabrics,
without knowing what the mineral
was.
Did the pseudomorphs grow after
either a deformation fabric or veins?
Do they contain inclusions, and are
these identifiable as minerals?
Are the pseudomorphs less deformed

than their host rock?
Are the pseudomorphs deformed, or
do they define a fabric?
Are pseudomorphs concentrated along
veins, contacts or faults?
Do they occur in reaction-zones?
Is it possible to determine the grade of
the pseudomorphing minerals?
If inclusions are visible, what is their
relationship to the host pseudomorph
(as Section 7.2.3)?

7.3.3 Pseudomorph identification

Many pseudomorphs have a chemical
composition very close to that of the
mineral they replace. That is to say, the
pseudomorphing involved minimal
reaction with the host rock, and generally resulted from a change in grade of
metamorphism. Compositional similarities can then be used as evidence of
identity of the minerals concerned.
Table 7.1 shows useful mineral information for common pseudomorphs.
*This kind of evidence does not apply in
reaction-zones or at contacts* between
siliceous and other rock-types.
Pseudomorphing there is usually
driven by the need to *destroy* the
previous chemical distributions and
create new ones.

7.3.4 Reactions and zoned pseudomorphs

If a marked compositional zonation
exists within the pseudomorph or the
margin of its host rock, then the whole
should be carefully drawn and (as far as
possible) annotated with mineral
names. If more than a quarter of the
pseudomorph is a rim of markedly different material from the core, compositions cannot be reliably used in identification. (Identity may still be clear from

situation or distinctive shape.) The
causes of zonation should not be interpreted in the field: there are too many
possibilities.

7.4 Veins and pegmatites

Veins and small intrusions of various
kinds (collectively called veins in this
chapter) may make up only a small
proportion of a rock unit, and give the
appearance of being of extraneous
origin, but they can be full of useful
information. Their material may:

1 Have been derived totally from out
  of the surrounding rock.
2 Have been brought in totally from
  great distances along fractures.
3 Result from equilibration of a
  foreign fluid with the local host
  rock, with partial transfer of
  material before crystallization.

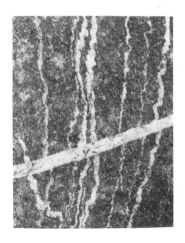

Fig. 7.15 Veins. Early calcite veins are
folded and cut by a later vein, in calcite
marble. Field of view: 7×9 cm.

These possibilities may not be distinguishable in the field. It may also not be clear whether the solid material crystallized from an aqueous (or other non-silicate) fluid, or whether it is a crystallized melt (particularly if it has suffered later deformation). A rock-mass description must record veins but avoid any interpretation.

7.4.1 Vein patterns

Veins are commonly concentrated in bands, and can be used to identify rock-bands, deformation bands (Section 6.4), grading and shear-directions. *Extension of competent rock-bands* can result in *cross-veins* between the pulled-apart sections. These initially develop at a high angle to the band, but may have been deformed later. They can highlight discrete bands (Fig. 7.16), grading (Fig. 7.17) and otherwise ill-defined layers in banded rocks (Fig. 4.5).

In shear-zones, extension veins initiate at about 45° to the zone boundaries, so that en echelon arrays are developed (Figs. 6.15 and 7.22). Such arrays highlight the existence of

Table 7.1 Common pseudomorph minerals

Minerals replaced	Common replacements (most common , in italics)	Less common replacements
Plagioclase Epidote Zeolites (Prehnite) (Lawsonite)	*Epidote group* Plagioclase Zeolites (sometimes + albite, or a white mica)	Lawsonite Pumpellyite Prehnite (Hydrogrossular in serpentinites)
Olivine	*Serpentine*	Talc Chlorite
Pyroxene	*Amphibole* Chlorite	
Amphibole Garnet	*Chlorite* Amphibole	Biotite
Biotite	Chlorite	Chlorite + sericite
Kyanite Andalusite Sillimanite	Sericite White mica	Kyanite Andalusite Sillimanite
Feldspar (K or Na) Topaz	—	White mica Sericite
Staurolite Cordierite Chloritoid	Chlorite + white mica Chlorite + sericite Biotite + white mica	Chlorite
Periclase	Brucite	
Blue amphibole Jadeitic pyroxene	A mat of fibrous green or grey–green amphibole.	

shear-zones where they might go un-
noticed, particularly if strain is very
low. In some cases a shear-zone is a
*rock-band of below average com-
petence* in a stratified sequence. Vein
orientation and comparison of rock
strains should distinguish between an
extended competent layer and a sheared
incompetent one. In other cases, shear-
zones do not correspond to rock-
bands, and may develop in *conjugate
sets* (Fig. 7.18). These usually cut
locally more competent units. Their
individual veins may vary greatly in
aspect ratio, and may be *straight or
sygmoidal* (Figs. 7.18 & 7.19). They lie
approximately perpendicular to any
local shear-zone fabric, having
opposite sense to the shear-direction
(Fig. 7.18). Shear-zones of veins can
exist on any scale from microscopic to
hundreds of metres width.

pattern is different from that in shear-
zones. Streaks of vein material precipi-
tate and get caught along the band
margin and between the foliae as they
kink over (Fig. 7.20).

Fig. 7.17 Cross-veins, near Fig. 7.16,
restricted to the upper half of a bed and wider
close to its top surface, showing it to be
graded and inverted. Bed width: 10 cm.

Fig. 7.16 Cross-veins highlighting a single
bed (otherwise not distinctive) in calcareous
slates.

Kink-bands can also have contem-
poraneous local veining, but the vein

Fig. 7.18 Shear-zones in sandstone, with en
echelon extensional veins and en echelon
dark pressure solution stripes. A conjugate
set, shortening in a nearly vertical direction.
Field of view: 60×90 cm. (S.W.Dyfed,
Wales)

Fig. 7.19 Irregular en echelon veins. Shear direction nearly vertical, sinistral displacement. Field of view: 17×25 cm.

7.4.2 Vein shapes, orientations and fibres

Wherever sheet veins or intrusions have *step-like offsets*, they should be identified as one of the four types shown in Fig. 7.21. Shapes to look out for at *vein ends* are splays with a threshold take-off angle (Fig. 7.22) and veins which pass into a radiating brush-like array of microscopic veinlets usually discernible only by weathering (see Fig. 2.3). Any irregularity of opposing *vein margins* may match across the vein, showing the over-all opening direction, provided this lies approximately within the plane of the exposure surface. Where *banding* is cut by veins, the sense of offset of its trace across a vein may be visible on a rock surface, but it may need recording on several surfaces to show the true opening

direction (which will otherwise remain ambiguous).

Vein shapes may result from post-veining deformation. In particular, veins may be the active element in folding (Section 6.5) or boudinage. Where strain is high, veins may become unidentifiable and effectively concordant with other lithological elements, and so contribute to banding and to lamination fabrics. Any inter-

Fig. 7.20 Veining simultaneous with kink-band development in slate. Cleavage in the vertical kink-band is horizontal. Elsewhere it runs from top right to bottom left. A vein runs across it, left to right, through the centre. For a short distance above the vein the dextral displacement has been accommodated by a vein off-shoot instead of kinking. Elsewhere along the band, white vein material has been caught here and there as thin streaks between the kinked flakes, and also along the kink-band boundaries (top). Field of view: 15×25 cm.

mediate deformation stage which suggests this should be recorded.

Veins which continue to develop during deformation may show relationships between their shapes and those of constituent mineral grains, if the veins widen by successive extension of their grains into fibres.

1 Continuous vein fibres can join the points on opposing margins which were originally in contact.
2 If the opening direction has changed during the history of the

vein's opening, this may be recorded by curved fibres, whether syntaxial or antitaxial (Fig. 7.22(h)).

3 Cross-veins of limited original length may eventually be extended by a distance along the rock-band much greater than their original length. Fibres within veins can show whether they have opened across or along their length (Fig. 7.22(g)).

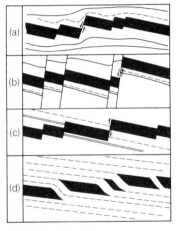

Fig. 7.21 Step-like off-sets of an intrusive sheet (in black; illustrated here as locally concordant with host-rock banding). (a) Oblique extension of a competent layer after intrusion. Active faulting of the layer is accommodated by ductile flow of the host-rock. (b) Passive faulting of the layer and its host-rock after intrusion. (c) Transform-style off-sets (initiated at the beginning of intrusion) of a sheet intruded oblique to host-rock banding on a larger scale. (d) Bridges of host-rock separating sections of a sheet intruded oblique to host-rock banding or fabric. Sections join up in three dimensions where bridges end.

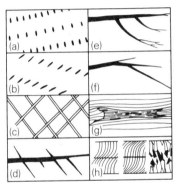

Fig. 7.22 Vein patterns, shapes and elements. (a) Extensional veins perpendicular to their bands, which have been lengthened. (b) En echelon extensional veins along shear-zones. (c) Veins intruding conjugate shear fractures, which may have initiated earlier. (d) Vein off-shoots picking out structure or fabric of the host-rock. (e) Vein splays near a termination. (f) Rotation of extension direction between successive propagation of branches of an extension vein. (g) Vein fibres (centre) along the length of a vein lying parallel to cleavage (trace indicated by lines around the vein). (A cross-vein with its original width extended to become far longer than its original length.) (h) Rotation of extension direction shown by syntaxial fibres (left) and antitaxial fibres (centre). Different mineral generations (right).

Fibrous or bearded pressure shadows or pressure fringes (Fig. 7.9) are small scale versions of these phenomena (Section 7.2).

7.4.3 Recording veins

In the field, record first any features similar to those mentioned or illustrated above. That is:

1 *The pattern of veins*, the orientation of this pattern, its size, and what it represents.
2 *Individual vein shapes*, dimensions and orientations.
3 *Opening directions*, and the type of evidence.

Annotated sketches are the best way to do this.

In structurally complex rocks, try to answer the following questions:

Are the veins folded (Fig. 7.15)?

Do veins cross-cut folds, or lie within fold structures in some rational orientation or position which shows the folds to be earlier?

Are veins and banding oblique to each other, and are both folded? If so, which is the active feature of the folding?

Is there shearing along the vein? If so, is there evidence to show whether the shear developed because of the existence of the vein, or whether the vein was located by the existence of the shear-zone?

Is strain concentrated at one side of the vein, or symmetrical about it? Is there concordance of any deformation fabric in the vein with a fabric outside (taking into account refraction, which is almost bound to occur from competence contrast with the host rock)?

Are there more complex geometrical relationships between vein and deformation, such as shown in Fig. 7.23?

Fig. 7.23 A complex vein pattern in slates created by continuing deformation and preferential vein growth along a lithological contact and coexisting deformation.

Then, *draw an example of grain shapes and textures* within the veins. They may be drusy, fibrous, comb-structured, etc. but should not be represented merely by such inadequate words. *Annotate* the sketch with the names of minerals (or brief descriptions if not identified) and then answer these further questions about minerals of the vein and its host rocks:

What are the *proportions of minerals*?
Can the same minerals be identified in the host rock?
Do proportions vary:

1 From the centre to the edge, gradationally, cyclicly or step-wise?
2 Along the vein? If so, does this show any correlation with the local host rock-type?

3 As several slightly cross-cutting bands, representing several episodes of veining?

Are there *early minerals* (or relics) in the vein, which may once have existed in the host rock also? (Draw shapes.)

Are there minerals in the vein which only occur as patchy *late growths* in the host rock?

Are there *accessory minerals*, of potential geochemical or economic interest, concentrated in the vein, or along its margins?

Does the host rock change on approaching the vein? If so:

1 Does the rock show gradual or successive sudden changes of mineral content on approaching the vein, suggesting a chemical reaction between the two? (See Fig. 7.24.) If so, what are the changes discernible within that portion which was host rock? Is there evidence of complementary alteration of the vein edge? Is there any separate reaction-zone

of new material separating the two? (See Chapter 8.)

2 Does the host rock develop at its margin an assemblage of different metamorphic grade? (See Section 10.1.) For example, a vein of greenschist minerals may be bordered by greenschist rock in an amphibolite or other basic host (Fig. 7.25). Such alterations are usually hydrating, and down grade.

Fig. 7.25 Greenschist alteration and segregation of host-rock by an albite vein. The host is a pale eclogitic rock of basic composition. Its margin is marked by mafic concentrations of actinolite crystals (dark). The complementary calc-aluminous portion of the basic composition has contributed to the vein, which is mostly blocky albite but contains some needles of epidote and a few flecks of chlorite. Field of view: 5×4 cm.

Fig. 7.24 Host-rock (calc-silicate schist) darkened by reaction at its margins with a pre-metamorphic vein.

3 If the change is only one of mineral proportions, does the rock become more like the vein in mineral

proportions nearer the contact, or does it become depleted in the vein minerals (suggesting that the vein derived, at least in part, from the local host rock)?

Contacts and reaction-zones

8.1 Igneous contacts—aureoles and metasomatism

Contacts between igneous rocks and others can be complicated. Their origins may be depositional, intrusive or faulted. If igneous material was deposited or intruded hot, the adjacent country rocks may have suffered a series of heating effects. Magma or lava may also have been affected by the proximity of a cool, static, country rock. Thereafter, the contact is liable to a retrograde sequence of reaction-zone effects as the complex cools. It then becomes liable to all the same normal reactions as any other contact (Section 8.2).

8.1.1 Intrusion geometry

The *shape* of the complete igneous body and of details of the contact should be considered. Small intrusions may have the same forms and patterns as veins (Figs. 7.21 & 7.22(a)–(e)). Details may include small ramifying side-intrusions, back-intrusions of mobilized country rock, and effects of deformation during or after igneous consolidation.

In the field, over-all shapes should be mapped out or sketched, according to size. Details of contacts should be sketched and annotated with mineral contents and proportions, which often change progressively in thin intrusive fingers of one rock-type into another.

8.1.2 Aureoles

Recognition: The direct affect of heat on country rock is either a localized *baked contact* of a few metres width or less, or a wider *aureole.* These rocks have suffered *contact metamorphism,* that is, recrystallization or the crystallization of new minerals as a result of increased temperature. This is most easily seen adjacent to shallow intrusions, or at the base of thick lava flows. Typical effects include:

1 *Bleaching* of quartzitic rocks (and *colour changes* more generally).
2 *Hornfels formation* from fine-grained rock-types.
3 Development of *spots* in certain rocks.
4 *Blurring* of the edges of previous large mineral grains or patches (amygdales, phenocrysts, single-mineral clasts, etc.), and of the textures of coarse crystalline rocks.

In the field, note what kind of features are diagnostic of the contact metamorphism, and how they change on a traverse towards the igneous contact, starting from beyond the aureole. If possible, a sequence of *aureole zones* should be defined *using the simplest characters available.* Such characters are *colour, grain-size, spacing of joints, type of fracture* when hit by a hammer, or the *minerals in veins* (e.g. prehnite then epidote, in basic or intermediate igneous country rocks). Any identifiable changes in minerals should be recorded, but it should be no surprise if these are rare additions to zoning which is better defined by the rocks' physical properties. Aureole zones should be mapped, if exposure is adequate. This may mean making a more detailed local map around an intrusion, if the general map is of unsuitable scale. Either in addition to, or in place of a map in very poorly exposed ground, an example transect should be constructed along a line of good exposure which traverses the aureole. This should record zone widths and variations within zones which are not part of the zone definitions.

8.1.3 Aureoles in regional metamorphic terrains

Igneous intrusion into deep metamorphic rocks, *already at moderate temperature* and grade, produces a broad 'aureole' of increased grade, which is only defined by minerals, rather than by changes of physical properties, and which may not follow the outlines of the igneous contact in detail. As such, it is unlikely to be shown up except by general mapping (if it can be distinguished at all from variations in the regional metamorphism).

Intrusion of igneous rocks into regionally metamorphosed rocks *after they have been uplifted* to shallow depth will superimpose a definable aureole onto pre-existing schists, etc. Hornfelses, visibly derived from medium or high grade schists, should be discernible at outcrop. Their loss of schistosity and tendency to irregular curved fracture may be taken as evidence of uplift after regional metamorphism and before intrusion.

The aureole of an intrusion occurring before regional metamorphism is unlikely to be recognizable, unless the latter is of very low or low grade. In this case, the hornfelsed rocks will generally constitute a more competent mass than their unhornfelsed equivalents, and this will be shown up by the structural relationships of a later fabric, and possibly by a difference in the type of cleavage.

8.1.4 Metasomatism

This name is given to chemical alteration of country rocks, with the assumption (now considered less than certain), that:

1 Metasomatizing fluids emanated from within the igneous intrusion.
2 Metasomatism occurred while the magmatic mass was hot.

Such alterations are highlighted by a *loss of some minerals* from the normal assemblage of the country rock, and the *addition* of a *metasomatic mineral* rich in a particular minor element (e.g. tourmaline, rich in boron). *Tourmaline, fluorite* and *topaz* are commonly metasomatic, but there are many other possible minerals, including a number of pale-coloured 'white micas'.

In the field, mineral occurrences may be declared metasomatic either if they are rich in 'metasomatic' minerals, or if chemical alterations are strongly localized along zones or around veins, near the igneous contact. (The latter implicates both the igneous body and fluid movement through zones or veins.) Sometimes such zones can be mapped out, or even zoned, by mineral contents and type textures. This should be done where possible. Otherwise, localities should be marked on the map, and the metasomatic rocks of each locality described. Over all, it is important to produce:

1 Sketches of the minerals and tex-

tures of the most extreme metasomatic rocks.

2 An ordered list of each country rock's minerals in their order of replacement by the metasomatic mineral (e.g. tourmaline replaces: hornblende, biotite, plagioclase, K-feldspar, quartz).

8.1.5 The margin of the igneous body

Three causes of compositional variations in the margins of intrusions need to be distinguished.

1 *Magmatic variations.* These may be the product either of more than one magma, if the intrusion is composite, or of crystal sorting. Sorting may be by grain-size or mineral species, and may result from layering at the base of a magma chamber or from shearing of a magma boundary layer next to a contact. There are also the possibilities of crystalline growth outwards from the wall of an intrusion, and of fine-grained chilled margins. All these variations are originally in the size and proportions of igneous minerals.

2 *Effects of hot circulating fluids.* These effects may be common, but are only likely to be discernible in the field if they involve changes which could not have occurred after cooling of the intrusion. The two most likely cases are *late growth of hornblende* in basic intrusions, and alterations of a *metasomatic* kind to the intrusive igneous rock (as distinct from the country rock). Zones of hornblende rock with gabbro texture, bordered by gabbro in which pyroxene is rimmed by hornblende (Fig. 8.1) indicate localized influx of aqueous fluids. In regions of only low grade metamorphism

(too low for metamorphic hornblende) this influx must have occurred while the intrusion was still hot. Metasomatic effects (e.g. Fig. 8.2), again likely to be localized in discrete zones, indicate considerable fluid flow, probably resulting from the heat of the intrusion. *These effects involve few metamorphic or metasomatic minerals, in zones cutting the igneous intrusion.*

3 *Back-reaction.* Reaction-zone effects are usually restricted to a few metres, or sometimes centimetres, from the contact, and they *vary with the type of country rock.* If the grade of reaction-zone minerals is greater than that of the country rock, reaction must have occurred before cooling of the intrusion. *These effects involve thorough major changes in minerals, sometimes in a series of zones, along the contact.* The considerations applying to reaction-zones generally are dealt with in Section 8.2.

In the field, variations in minerals near the margin of an intrusion should be looked for, and categorized as above on the basis of *minerals, textures* and *distributions.* Textures should be drawn, with an indication, perhaps as another sketch, of the variation of texture with location over distances of a few metres.

8.2 Reaction-zones and chemical changes at contacts

Chemical reaction between adjacent rocks subjected to metamorphism is both common and commonly overlooked. Grossly abnormal compositions, easily identified, are usually restricted to a few metres, or less, from contacts. Less easily visible changes may extend further. As these changes are likely to be largely effected through fluids, and contacts can provide pre-ferred paths for fluid movement, it may be difficult to distinguish some reaction effects from hydrothermal ones. The processes can be complementary.

Fig. 8.1 Hornblendic zones in gabbro. A strip of black and white rock exists along the left side of the picture, and just connects (at about ¾ of the picture height) to a similar patch on the right side. These patches are of hornblende and white plagioclase, but have gabbro texture. The unaltered gabbro (pyroxene and plagioclase) is duller (centre of lower half and top centre edge).

8.2.1 Recognition of reaction

In the field, the first task is to recognize reaction. In simple cases this is shown either by *the loss of one or more minerals* from the assemblage which prevails further from the contact, or by

a *change in composition of one or more minerals* (possibly visible as a change in colour or size). Only more locally at the contact, and with extreme instabil-ity between the two rock-types, will new minerals be found. Because these changes are generally small, their exist-ence can only be demonstrated by showing their restricted location, with closer correspondence to the vicinity of the contact than to compositional trends in the rocks on each side.

Fig. 8.2 Radiating needles of tourmaline pseudomorphing a potash feldspar pheno-cryst in a granite. (Cornwall)

Whether or not mineral changes are visible, a record should be made of any general features which suggest that adjacent rocks have affected each other. Bleaching or blackening of the rock (Fig. 7.24), or surface staining, or different resistance to erosion along a contact all suggest reaction. These should be reported, as also should any evidence for alternative causes of alteration, such as mineralization or localized cataclasis.

8.2.2 Reaction-zones

Where extreme reaction has occurred, there are potentially three distinguishable bands of reaction-zone rocks. On either side there are rocks which despite chemical alteration retain features of their previous rock-types. Between these there may be a central band of rock which lacks any previous structure. This may be entirely new material, precipitated during the reaction, but it may also contain earlier rocks which have been so altered as to destroy any trace of their original character (Fig. 8.3).

Within the bands on each side, alteration is likely to increase gradationally towards the contact. Any new mineral is likely to be seen first as a slight alteration, then as a pseudomorphic replacement, of one or two minerals of the original rock. On approaching the contact, the new mineral may successively replace more minerals of the earlier assemblage. Such replacements are often well displayed by metasomatic minerals and by back-reaction replacements of igneous textures at intrusive contacts (Sections 8.1.4 & 8.1.5).

In the field, the textures of different stages of alteration should be sketched, and a record made of whether they define a regular series of zones, or whether replacement increases patchily towards joints, compositional bands or foliation planes which might have been channels for fluid transport. If zones are regular, their widths should be measured, perhaps as a transect. Otherwise, an example of their irregularity should be sketched or grid-mapped. If several new minerals appear as the contact is approached, their order

should be recorded in different places and particularly in different original rock-types. The consistency of their order, or of correlation between order and original rock-type, should be recorded.

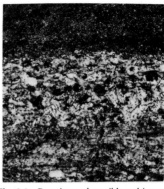

Fig. 8.3 Reaction and possible melting at a contact in migmatites. The rocks are now amphibolite (top) and arkosic quartzite (bottom), both fairly fine-grained, with a strong deformation fabric. There is a 3.5 cm band (centre) of coarser dioritic rock with a weak fabric, presumably formed by reaction between basic and sedimentary rock compositions, perhaps with melting. Elsewhere, this dioritic rock-type appears to intrude the amphibolite.

Within the central zone of new material, relationships are usually simple. The number of minerals is normally less than outside, and changes are usually regular (unless complicated by later deformation). If several bands occur, having the same few minerals and similar texture, they are successive generations, and may have slight cross-cutting geometry. Only with the most extreme differences in rock-type are there likely to be several zones having different mineral assemblages (the best known examples of which are at contacts of either ultramafic or carbonate rocks with siliceous rocks). *Straight-forward reaction between adjacent masses produces an unreversed simple sequence of zones each of fewer minerals than the rocks at the sides. Any case of very many bands, of reversals of the sequence, or of irregular accumulations of a number of minerals is dominantly of vein or hydrothermal origin, not a reaction-zone.*

In the field, the minerals, textures and zone widths should be recorded. An annotated sketch is best. It is more important to collect and report this evidence than to decide between reaction-zone and another kind of deposit, if this is in doubt.

8.2.3 Deformation of reaction-zones

Reaction-zones are bands of rock which differ greatly in their physical properties, the banding constituting an over-all anisotropy, in between two ad jacent masses which probably differ i competence. In any subsequent de formation, reaction-zones are liable t be strongly disrupted, producin features such as isoclinal folds, boudin and blocks of monomineralic or b mineralic material, lobed contacts, an 'tectonic intrusions' (Fig. 8.4). *In th field*, keep a look out for disrupte zones of a few simple materials, an draw any features which are recog nized.

8.2.4 Metamorphic grade and age

Reaction-zones develop only whe two dissimilar rocks are adjacent t each other at certain specific condition of metamorphism. Sometimes, reactiv rocks are physically put into contact by faulting, or by intrusion an crystallization of magma, at suitabl metamorphic conditions. Sometimes

Fig. 8.4 Tectonic intrusive dykes. This view is of the top surface of a meta-lava formation, exposed by erosion of overlying serpentine and talc schists. Deep (2 m) parallel gashes, filled with actinolite–talc and actinolite–chlorite schists, cut down into the meta-lavas. (One metre rule.)

potentially reactive rocks are deposited together at the earth's surface, but no significant reaction occurs until they are brought to higher pressures and temperatures at which metamorphism then takes place. Sometimes, rocks are genuinely stable together at certain conditions of pressure and temperature, but react at others. For example, olivine-rich bands and plagioclase-pyroxene bands of a layered gabbro are stable together at high temperatures, such as those of igneous crystallization, where they are part of an over-all olivine + pyroxene + plagioclase assemblage. At low metamorphic grade, they metamorphose to two different rock-types (serpentine and greenschist respectively) which are not part of the same chemical equilibrium. At their contacts, they react to form a reaction-zone of chlorite + tremolite. Grade is vital in determining whether reaction between two rocks will occur, and what reaction-zones will be produced.

Metamorphosed rocks do not always equilibrate internally to later conditions of metamorphism, particularly to those of a lower grade, but if reactions occur at contacts under such conditions, the reaction-zone will record the lower grade. Grade discrepancies between reaction-zones and neighbouring rocks may be useful in establishing a history of grades and displacements of rock-masses. Where major thrusting or igneous activity has placed rocks together, the rocks each side of the reaction-zone may themselves differ in grade. It is even possible that a thermal gradient existed between the rocks each side of the contact as reaction between them occurred. For all these reasons, the grade of a reaction-zone should be considered and it should be reported whether or not there is any obvious difference between its grade and that of either of its neighbouring rocks. If the zone is of lower grade, there may be other lower grade features (shear-zones, perhaps) with which it may be correlated, or it may display a late fabric. Any case of a reaction-zone of higher grade, or with grade changing from band to band through the zone, is extremely unusual, and could be useful in geotectonic interpretations. Such cases should be highlighted, and described as fully as possible.

9

Faults, mylonites and cataclasites

In metamorphic terrains, displaced but relatively undeformed blocks of rock may be separated either by discrete fault planes or by deformed zones of finite thickness. The usual care is needed in the field to notice and record proved and potential lines of displacement of markers and contacts.

Extra care must be taken to watch out for possible pre-deformation faults in deformed areas. Within deformed zones, there is a wide range of potential rock-types, corresponding to the whole variety of physical environments in which displacements may occur. Finding and identifying fault-rocks should help to locate otherwise overlooked faults, and known displacements indicate localities where fault-rocks may exist.

9.1 Faults

9.1.1 Shear fibres

Detailed irregularities on a fault plane lead to areas of mineral removal and areas of mineral regrowth. Mineral grains so grown can be extended by the continuing motion to produce fibres running almost along the surface in the *displacement direction*. This can be very valuable in determining the fault type (normal, transcurrent, etc.) where only the dip separations of layers can be ascertained by comparing the two sides. However, because such processes are common at very shallow depth, the displacements recorded may be later than anything tectonically significant. *In the field,* the existence, mineral species and direction and sense of movement of fibres should be recorded, and their pattern should be sorted out, to see whether it is of tectonic significance. (Fibres often show only that large masses of rock are creeping down present-day hillsides.)

9.1.2 Chemical changes near faults

Faults provide local physical stresses, a zone of fluid movement, and a juxtaposition of unlike rocks. Chemical processes are therefore likely to affect the wall rocks to some distance each side. *In the field,* look out for cloudy alterations of feldspars, reddening from oxidation and precipitation of ferric oxides and hydroxides, carbonate replacement of igneous rocks, and dull chloritic alterations to schists. Even where chemical effects are not noticeable, there may be changes in weathering, or in the character of joints, which should be recorded.

9.2 Fault and shear-zone rock-types

Faults are commonly sites of deposition of *veins, sedimentary dykes* or *fluidized sediment,* and are often *mineralized* (see Section 7.4). *Look out for such materials,* particularly in clearly defined fault-zones, *as well as for fault-rocks proper,* which are produced predominantly by physical breakdown of the wall rocks and may not be confined so clearly to precise zones.

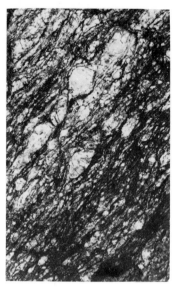

Fig. 9.1 Mylonites. Dark mylonitic streaks and cracked porphyroclasts (the largest 1 cm across) of quartz and feldspar in a mylonitized granite. This rock is partly 'mylonite' and partly 'protomylonite', according to the subdivisions in Table 9.1.

Displacement between two bodies of rock may be accomplished through any situation between fault motion on a single plane, and shear across a zone so wide that effects on the rocks in the zone are negligible. *Shear-zones* containing normal rock-types, either similar to or different from those outside, are dealt with in Section 6.4.4.

For the special rock-types of faults and shear-zones ('*Fault rocks*'), the scheme set out in Table 9.1 is similar to several which have been published, and

combines distinctions visible in the field to produce what is thought to be a genetically significant classification. It differs from classifications produced before the 1970s (before the role of intracrystalline plasticity in rock deformation had been recognized) which were based on the previous concepts of 'cataclasis' and 'crystalloblasis'. It also omits terms ('blastomylonite', in particular) which some geologists say require microscopic distinction between modes of 'recrystallization'.

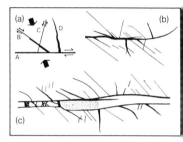

Fig. 9.2 Pseudotachylyte geometry. Dextral principal shear illustrated here. (a) Principal elements: principal shear fracture (A), often along banding; minor shear fractures (B) of the same sense; minor shear fractures (C) of opposite sense and at a high angle to A; extension fractures (D). B and C are a conjugate set. (See Fig. 6.15.) Pseudotachylyte is found along A, D and some of B. (b) A simple fracture pattern, with one principal shear fracture and showing a termination. (c) A case of paired principal fractures. Minor fractures are more densely situated between these than outside them. Intersecting fractures can separate off 'clasts', with somewhat rounded-off corners and edges. In places, dense fractures merge into a full pseudotachylyte breccia (indicated schematically by stipple) in which the matrix is entirely pseudotachylyte.

As shown in Table 9.1, there are two series, *cataclasite* and *mylonite* in their

Table 9.1 Field names for fault-rocks

	NO VISIBLE FABRIC	VISIBLE DIRECTIONAL FABRIC
Glassy Like volcanic glass or devitrified glass	*Pseudotachylyte* (Record geometric features, as in Fig. 9.2.)	
Incohesive Visible fragments by bulk:	*Fault breccia and gouge*	
< 30% (Gouge)	Fault gouge	Mylonitic fault gouge
> 30% (Breccia)	Fault breccia	Mylonitic fault breccia
Cohesive non-glassy *Fine matrix + fragments* Fragment size: > 5 mm	*Crush breccia* Crush breccia	Mylonitic crush breccia
90–100% — 1 mm–5 mm (Fine)	Fine crush breccia	Mylonitic fine crush breccia
< 1 mm (Micro)	Crush microbreccia	Mylonitic crush microbreccia
50–90%	*Cataclasite* Protocataclasite	*Mylonite* Protomylonite
10–50%	Cataclasite	Mylonite
< 10%	Ultracataclasite	Ultramylonite
Visible new matrix grains		'Mylonitic' schist or *gneiss*

(left vertical label: Proportion of visible fragments)

broader senses, which are visually distinct. Within each series, the distinctions reflect the degree of grain-size reduction. Between the two series is probably a difference in process of grain-size reduction. Members of the same series are likely to occur in close proximity. If mixtures of the two series are found, such *crush melanges* should be noted with, if at all possible, the relative ages of the two component types.

Folding is common in mylonites, mylonitic schists and mylonitic gneisses, and sometimes has *sheath-fold* form. Fold shapes should be sketched, particularly on a number of differently oriented exposure surfaces, and the three-dimensional fold shapes elucidated.

A *mylonitic schist* or *gneiss* (e.g. Fig. 9.3) has had its grain-size reduced towards the fineness required for continuing ductility at the physical conditions of the particular deformation. A banded rock within a coarser host should be termed 'mylonitic' if it can be shown to occur in a definable deformation zone, or if it contains *porphyroclasts.*

The greatest identification problems are usually with the *fine-grained rocks of the mylonite series* having acid or intermediate plutonic or gneissic parentage. Some are slates or phyllites. It is very easy to overestimate the amount of sheet-silicate and to underestimate the amount of fine-grained quartz and feldspar in such rocks (see Fig. 9.4). Others are chert-like. Those with scattered and rounded-off broken grains of quartz and feldspar can appear meta-volcanic or volcaniclastic. Granitic rocks with only protomylonitic alteration of the edges and corners of grains can be mistaken for arkosic metasediments. If these contain chunks of rock which have escaped alteration, they may appear to be meta-conglomerates. Many of these awkward rocks may be given away by a rather unusual greyish look, but this cannot be relied upon. Once a mylonite zone has been identified, the association of several such rock-types within a mappable unit is fairly easy to follow.

Fig. 9.3 A mylonitic gneiss (meta-granite)

Not every zone of mylonitic rocks is a zone of simple shear. The roots of 'nipped-in' synclines are zones of intense deformation of basement rocks due at least in part to competence contrast with a less competent cover. These zones are likely locations for both slithers of highly deformed (possibly mylonite) meta-sediments and of mylonitic basement rocks which look like meta-sediments. It may be necessary to map such zones, without knowing at all exposures the exact rock-types and their origins.

In the field, fault-rocks found should be categorized as one of the six major types of Table 9.1. If the sub-category is also obvious, then its more detailed name should be used. Notes should identify the minerals of porphyroclasts and in the case of 'mylonitic' rock, the matrix, and also the physical properties of fine-grained rocks or portions, and the shapes and orientations of chunks, bands, fabrics, etc. If a normal rock type name is applicable, this can be

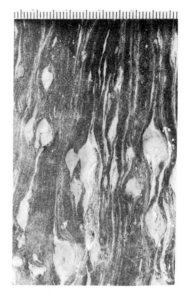

Fig. 9.4 Two views of the same mylonite. A section (top) cutting the fabric shows up fragments of previous rock in a dark matrix. A fracture along the mylonite fabric (bottom) shows mostly new fine-grained phyllitic material, which appears paler here than in cross-section.

Fig. 9.5 Competent dolomite lumps being streaked out into a calcite mylonite (or mylonitic marble).

combined with that of the fault-rock type (e.g. mylonitic actinolite schist, green slaty mylonite).

Cross-cutting of one fault-rock type by another should be either sketched or grid-mapped, according to scale, and if commonly found, a clear statement should be made about the degree of consistency from one exposure to another of the relative age of different types.

There may be a change in fault-rocks on approaching the wall rocks to either side of the zone, and, where the zone is wide, there may be fault-rocks of identifiably different parentage. Such changes across the zone should either be mapped or logged.

Reference tables and checklists

10.1 Compositional categories and their grade indicators

A Ultramafic

Diagnostics: Rocks rich in some of: *serpentine, talc, olivine, anthophyllite, cummingtonite, enstatite, bronzite*. Including *SERPENTINES* and *PERI-DOTITES*.

Derivation: Ultramafic igneous rocks. Very rarely from associated very local and very immature sediments from an ultramafic source. *Often serpentinized before metamorphism.*

Mineral origins: Bronzite, augite and *chromite* are possible igneous remnants, while *olivine, plagioclase, spinel, garnet* and *magnetite* may be either igneous or metamorphic. Remnant igneous minerals need not accord with metamorphic grade. In addition to the diagnostic minerals and those mentioned above, the following metamorphic minerals may occur: *magnesite, brucite, chlorite, diopside, tremolite, phlogopite*.

Grade: Good rocks for indicating grade category. Serpentinites are *VLG* or *LG*. Peridotites are *MG, HG, VHG* or igneous. Abundant *talc* is usually *MG*. Abundant *anthophyllite, cummingtonite* or *enstatite* is *HG*.

For good definition, search out magnesitic or talcose exposures, or siliceous contacts, and look for these combinations:

Serpentine+quartz (without talc)	—*VLG*
Serpentine+talc (any *quartz* is *talc-coated*)	—*LG*
Olivine+talc	—*MG*
(Anthophyllite or *cummingtonite)+(olivine* or *talc)*	—just *HG*
(Anthophyllite or *cummingtonite)+(enstatite* or *quartz)*	—*HG*
Periclase; Spinel; Sapphirine	—*VHG*

Subdivision of *LG* and *MG* is possible by these combinations:

Serpentine+brucite	—*VLG* or *lower LG*
Serpentine+olivine	—*upper LG*
Olivine+talc+magnesite	—*lower MG*
Olivine+talc+ scattered *anthophyllite* or *enstatite*	—*upper MG*

Chlorite, phlogopite, tremolite and *diopside* are not good grade indicators in ultramafic rocks.

B   Basic and intermediate

Diagnostics: Rocks having a *mafic* and a *calc-aluminous* portion, and sometimes also a *quartz* portion, and a *biotite* portion. Including *GREENSCHISTS* and *AMPHIBOLITES.*

Mafic portion = Some of: *chlorite, amphiboles, pyroxenes, garnet.*

Calc-aluminous portion = Some of: *plagioclase, albite, epidote group, zeolites, prehnite, pumpellyite, lawsonite.*

Derivation: Basic or intermediate igneous rocks. Sometimes immature sediments from such a source. Widespread *carbonate* may occur in sediments or lavas subjected to hydrothermal alteration. Extremely rarely (if *MG* or higher) from dolo-marls (see **G** and Section 5.5.6). See Section 5.5.3.

Grade: These rocks give a broad indication of grade category, as shown in Table 10.1 (p. 103). However, changes at category boundaries may be gradational, and no distinction is possible between *MG* and *HG*. Biotite can occur at *LG. MG* and *HG*.

Garnet or *diopside* may be usable for defining zones in the field (using either first appearances or mineral proportions) but not as a universally reliable grade indicator. *Cummingtonite* may occur in hornblende-rich rocks. Igneous minerals may remain and need not accord wth metamorphic grade.

VLG: A colour sequence of *white* (zeolites, analcite) to *green* (prehnite, pumpellyite) to *yellow-green* (epidote), in fine-grained igneous materials, is prograde, within *VLG,* but may result from heating in a lava pile, not from later metamorphism.

C   Acid

Diagnostics: Much *quartz* and *K-feldspar,* and sometimes *albite* (or *plagioclase*), sometimes *mica.* Can include some *epidote, chlorite, amphibole, zeolite, prehnite* or *stilpnomelane.*

If too quartz-rich for an igneous rocks: see **D.**

If too micaceous for an igneous rock: see **E.**

Derivation: Acid igneous rocks. Sometimes from granitic conglomerates. Sometimes from arkosic sands and granule-conglomerates. A coarse igneous or sedimentary texture is normally retained, unless deformation is intense.

Grade: Almost useless for indicating grade. If present:

Chlorite; epidote	— *VLG* or *LG*
Stilpnomelane	— *VGL/LG* boundary
Biotite	— *LG, MG* or *HG*
MIGMATITES	— *HG*
Hypersthene with *garnet, cordierite, sillimanite;* or Sapphirine	— *VHG*

D   Quartzite

Diagnostics: Dominated by quartz. *QUARTZITES,* including *ARKOSIC* and *MICACEOUS* varieties.

Derivation: mature sorted sandstones and conglomerates. (Hence sometimes called 'psammite' = arenite.)

Grade: Generally useless for indicating grade, except:
White mica	— *VLG, LG* or *MG*
Al-silicate	— *HG* or *VHG*

(See **E**. Note presence or absence of K-feldspar.)

E   Semi-pelite and pelite

Diagnostics: Rich in *sericite* or *mica*, or containing an *aluminous mineral*. Other common minerals: *quartz, chlorite, feldspars, garnet*.

  Aluminous minerals = *chloritoid, staurolite, cordierite* and *Al-silicates*.

  Al-silicates = *andalusite, kyanite, sillimanite*.

  Including most *MICASCHISTS* and *SLATES*. Also some gneisses.

Derivation: A wide range of clastic sediments having a high proportion of fine-grained material. Broadly speaking: greywackes, siltstones and mudstones. ('Pelite' = argillite.)

Grade: Although showing a number of changes with increasing grade, category boundaries can be ill-defined unless either pelites proper (**E** +) are present, or *HG* has been attained. Note that *biotite* normally appears at some point within *lower LG*, and that *quartz* and *albite* may exist at all grades. Any *stilpnomelane* indicates *VLG/LG boundary*. At *VHG*, K-feldspar (and sometimes *quartz*) *may be lacking*, indicating loss of a potassic fraction, probably as partial melt. At *LG* and *MG*, K-feldspar can occur only in the absence of aluminous minerals (i.e., in semi-pelites not pelites).

  Some minerals in these rocks are PRESSURE INDICATORS:
Garnet	— intermediate or high P
Cordierite	— intermediate or low P
Kyanite	— high P. (Extreme high P if *HG*)
Andalusite	— intermediate or low P. (Low P if *HG*)

Further subdivision is possible for pelites proper, as below.

E +   Pelite subgroup

Diagnostics: Chloritoid; Staurolite; or the combination: *Aluminous mineral + white mica + quartz*.

Derivation: Chemically mature mudstones.

Grade: These rocks provide a subdivision of *LG* and *MG* and the diagnostics given above indicate this grade range. In addition to pressure indicators given under **E**:

  Indicators of *LG*: Chloritoid; Garnet + chlorite; restriction of any *kyanite* or *andalusite* to biotite-free rocks. Within *LG*, garnet indicates *upper LG*.

  Indicators of *MG*: Staurolite; Cordierite (+ *white mica + quartz*); Al-silicate + biotite (+ *white mica + quartz*). Within *MG*:
Staurolite + chlorite	— *lower MG*
Al-silicate + biotite	— *upper MG*; or low P; or both
Sillimanite; Garnet + cordierite	— *upper MG* at intermediate P
Staurolite	— Intermediate or high P; or low P at lowest *MG*

F Calcareous

Diagnostics: Rocks rich in carbonate minerals. *MARBLES.*

Derivation: Limestones.

    (This does not deal with hydrothermal or metasomatic deposits, with magnesite in association with evaporites or serpentines, or with sideritic sedimentary or meta-sedimentary rocks.)

Minerals in marbles: Pure calcite limestones metamorphose to marbles but do not change with metamorphic grade. Non-carbonate minerals in marbles usually represent detrital or evaporitic impurities *(quartz, chlorite, sericite, mica, gypsum)* which in larger quantities generate calc-silicates (category **G**, below). It is dolomites with low original clay content but a substantial amount of silica (normally chert) which metamorphose to Mg-silicate-bearing calcite marbles.

Grade: Marbles containing calc-aluminous or magnesian silicates can be good grade indicators, but their treatment is not easy in the field. Metamorphosed siliceous dolomites in particular are of somewhat restricted occurrences, and their minerals are not always readily identifiable.

Shaly or slaty fine-grained marbles indicate *VLG.*

Chloritic or micaceous marbles indicate *VLG* or *LG.*

Calc-aluminous minerals in marbles — as in **G**, below.

    The grade indicators mentioned below occur in metamorphosed siliceous dolomites (now calcite-rich marbles), see Section 5.5.5. The sequence of first appearances: *talc, tremolite, diopside, olivine* spans *LG* and *MG*, but it is possible for *talc* and sometimes also *tremolite* to be by-passed. *Chlorite* is not a reliable grade indicator in such rocks. The following minerals indicate the grade shown, *if coexisting with calcite* (but their absence does not refute such a grade):

Talc; Phlogopite; Dolomite+quartz	— *VLG* or *LG*
Tremolite; Actinolite	— *MG*
Wollastonite; Periclase; Spinel	— *HG*
Olivine; Dolomite+diopside; Scapolite	— *MG* or *HG*

Pseudomorphs: Brucite as periclase; *Serpentine* as olivine.

    At shallow igneous contacts, a series of zones of unusual minerals may be discernible. The first of these minerals are often *monticellite* and *melilite*, but identification is not usually possible in the field.

G Meta-marls and calc-silicates

Diagnostics: Either rich in both *calcite* and *silicates*; or rich in Ca-silicates, such as *diopside, plagioclase, grossular, epidote, wollastonite.* See Sections 5.5.5, 5.5.6 & 5.5.7. Are silicates as in **F** above?

Derivation: Sedimentary mixtures of carbonate and clastic materials.

Grade: Calc-silicates are theoretically good grade indicators, but in practice there may be difficulties with the mineral identifications needed. The major reactions in them do not coincide with grade category boundaries.

    VLG: a mixture of carbonates and of the silicates of *VLG* meta-clastic rocks (*chlorite, sericite, quartz*), normally as a calcareous slate.

    LG: a mixture of carbonates and of silicates found in various rocks at this grade (*white mica, chlorite, quartz, albite, epidote, actinolite, biotite*).

Upper LG, MG and *HG:* very variable assemblages, commonly containing *diopside.* Each of *biotite, actinolite, diopside* and *garnet* can persist throughout these grades (particularly in patches where all *calcite* has been used up). Diopside amphibolites may occur (Section 5.5.6).

The following grades are indicated by *coexistence with calcite:*

Chloritoid	*—lower LG*
Zoisite; Clinozoisite	*—LG* or *lower MG*
Plagioclase + white mica	*—upper LG* or *lower MG*
K-feldspar + plagioclase + quartz	*—upper MG*
Grossular + anorthite + quartz	*—MG* at low P; *HG* at high P
Wollastonite; Spinel	*—HG*

H   Rocks of high pressure of all compositional types

Diagnostics: Blue amphiboles; Green (jade) pyroxenes; Greenish or *brownish phengite micas; Lawsonite; Aragonite.* Normal chemical categories of minerals do not apply. Blue amphiboles combine NaAl with a mafic constituent. Jade pyroxenes contain NaAl in solid solution with either CaMg of diopside, or NaFe of aegirine, or both. Phengites are mafic rather than aluminous, having compositions similar to $(K-feldspar+biotite)$, $(K-feldspar+chlorite)$ or $(K-feldspar+orthopyroxene)$. Therefore, the normal compositional categories of rocks (A–G) are unsuitable for their subdivision.

Ha.   Marginal blueschist association

Blueschists or *blue amphibole-bearing greenschists* (usually fine-grained and of basic igneous origin) *amongst ordinary VLG or LG rocks.* A little *blue amphibole* may occur in metamorphosed plutonic rocks, marbles and meta-clastic rocks.

Blue amphibole may exist with *albite, chlorite, phengite, stilpnomelane, calcite, quartz, sphene, actinolite, epidote, lawsonite.* If CaAl-silicates can be identified, they indicate:

Lawsonite	—marginal to *VLG*
Zoisite; Clinozoisite	—marginal to *LG*

Hb.   Jadeite blueschist association

Blue amphiboles are *widespread,* with concentrations in meta-basic rocks. *Jadeitic pyroxene* occurs with *quartz* in acid, and some basic, and some meta-sedimentary rocks. *No albite.* Phengite in acid and pelitic rocks. *Lawsonite* in some basic and intermediate rocks and some marbles. *Aragonite* may occur, but may have reverted to *calcite.*

Other minerals: *stilpnomelane, garnet, sphene, actinolite, quartz.*

Metamorphism: *VLG. High pressure.* Abnormally cold geothermal gradient.

Hc.   Glaucophane eclogite association

Blue amphibole, omphacite, garnet, phengite, white mica (Na-mica), *carbonates, quartz, rutile* can all occur together, in rocks of many compositions, including basic and metasedimentary types. Basic rocks also contain *zoisite, clinozoisite,* sometimes *green epidote* and sometimes *green amphibole. Phengite, jadeitic pyroxene, quartz* and *garnet* may occur in acid and sedimentary compositions. *Blue amphibole, chloritoid, epidote* in some schists. Other combinations of the minerals cited.

Hc+. Even more extreme assemblages (alterating to glaucophane eclogites): *talc* (to amphiboles), *kyanite* (to Na-white mica), *chloritoid* (to white mica), *lawsonite* (to zoisite, clinozoisite), *omphacite, garnet, phengite, quartz*.

Metamorphism: *VLG* to *LG*. *Ultra-high pressure*. Abnormally deep, and abnormally cold geothermal gradient.

Hd. Eclogite association

Basic rocks of essentially *omphacite+garnet ('Eclogite')*, plus other minerals in small to moderate amounts, plus *rutile*. Other minerals can indicate variations of basic composition, e.g. *clinozoisite* where originally plagioclase-rich. Otherwise they belong to normal metamorphic rocks and may indicate grade of metamorphism:

Albite+clinozoisite+actinolite+white mica (Na-mica)	—LG
Hornblende+clinozoisite+white mica	—MG
Hornblende+kyanite	—MG or HG

Such rocks have equilibrated at high pressure, at the grade indicated by additional minerals, but with insufficient water for normal basic rocks (**B**). (Such a predicament is not applicable in the case of *VHG* 'granulites', acid rocks, or meta-sedimentary rocks, for which normal metamorphism includes anhydrous assemblages.)

Warning. Some eclogites contain *olivine* plus minerals of *HG* or *VHG*, and are probably mantle rocks, not produced by metamorphism of crustal material. In others, normal hydrous minerals occur, but result from late hydrous alteration, and are out of equilibrium with the eclogite.

Table 10.1

Summary table: main mineral constituents of some compositional categories at different grades

Grade category	MATERIAL			
	A Ultramafic	*B Basic igneous:*		*E Semi-pelite and pelite*
		Mafic portion	*Calc-aluminous portion*	
VLG	Serpentine (Quartz, magnesite)	Clays, chlorite (igneous relics)	Zeolite, pumpellyite, epidote, albite	Clays, chlorite, sericite, quartz
LG	Serpentine (Talc, magnesite)	Chlorite, actinolite (Garnet)	Epidote, albite	White mica, chlorite, quartz, biotite (Garnet, Al-minerals)
MG	Olivine, talc (Magnesite, anthophyllite)	Hornblende (Diopside, garnet)	Plagioclase	White mica, biotite, quartz (Garnet, Al-minerals)
HG	Olivine, anthophyllite, cummingtonite, enstatite	Hornblende (Diopside, garnet)	Plagioclase	K-feldspar, biotite, quartz, Al-minerals, (Garnet) —or migmatites—
VHG	Olivine, enstatite	Hypersthene, diopside (Hornblende)	Plagioclase	Hypersthene+Al-minerals (K-feldspar, quartz) or Sapphirine+other minerals

10.2 Minerals

A list of selected metamorphic minerals and their usual properties.

10.2.1 Main mineral groups and other common bulk minerals

Mineral	Hardness (Mohs' scale)	Description	Compositional Category (Section 10.1)
Quartz	7	*Glassy. Colourless*, except for a purple or blue–grey hue at *VHG*. Not subject to alteration. Trigonal.	**All**
Carbonates *Rhombohedral:*	4 (except Calcite 3)	Rhombohedral cleavages. Trigonal.	

	Colour:	Reaction with *cold* dilute hydrochloric acid:	Weathering:	
Calcite	White	Effervesces	Clean. (Rusty if ferroan.)	**BFG**
Siderite	Brown	Reacts quietly	Rusty stain.	—
Magnesite Dolomite Ankerite	White to yellow to brown	Reacts only if powdered.	(Degree of ochre staining varies with Fe : Mg ratio.)	**AFG**

Orthorhombic: Aragonite		*White. One cleavage, Effervesces* with cold, dilute hydrochloric acid.	H
Feldspars	6	*Grey, white* or *cream. Equidimensional.* Two cleavages at about 90°. Triclinic. Often difficult or impossible to distinguish one metamorphic feldspar from another, as twinning and clear grain-shapes are rare.	
K-feldspar		Sometimes yellow–pink (*HG*) or green–brown (*VHG*).	**CDEG**
Albite		Can develop clean, angular porphyroblasts (white or a plastic-like blue–grey) even at *LG*.	**BCDEG**
Plagioclase		Sometimes as inclusion-filled ovoid porphyroblasts in schists (as cordierite). *Igneous* feldspars are common remnants, and may show distinctive shapes and twins. *All* feldspars may be reddened, e.g. near faults.	**ABCG**

Pyroxenes	5–6	*Two equivalent cleavages, at 90°, along* grain length.	
Clino (Monoclinic):			
Diopside		White, grey or pale green. Ovoid, or stubby prisms. (Igneous relics, augitic or diopsidic, are common in meta-gabbros.) Aggregates glassy.	**ABFG**
Jadeite		Grey to pale *jade green*, as aggregates or needles. Fibrous or glassy.	**H**
Omphacite		Deep green. As aggregates or long prisms.	**H**
Aegirine-Jadeite		Transitional to jadeite.	
Hedenbergite		Dark brown or dark green. In iron-rich rocks and skarns.	
Ortho (Orthorhombic):			
Enstatite		Grey, or green. (Igneous *bronzite* : bronze.)	**A**
Hypersthene		Brown.	**BCE**

Amphiboles	5–6	*Two equivalent cleavages, at 125°, along* grain length.	
Clino:			
Hornblende ⎫ *Cummingtonite* ⎬		Deep green, brown or black. Short blades or squat prisms.	**ABCG**
Tremolite		Grey or green blades or needles. Sometimes feathery or fibrous.	**AFG**
Actinolite		Deep green or green–black. Shapes as tremolite. *Asbestos* will powder if crushed.	**BCFG**
Blue-amphiboles		*Glaucophane* is blue. Others are blue–purple, violet, or purply black. Shapes as tremolite. Colour can be partially masked by rims altered to green amphibole. *Blue asbestos:* DO NOT ATTEMPT TO POWDER—Dangerous if inhaled.	**H**
Ortho:			
Anthophyllite		Grey or pale green. Fibres, needles or blades, often radiating.	**A**

Epidote Group	6	*Prismatic, with one cleavage along length.*	
Epidote		Straw to *epidote green*. ⎫ Parallelogram sections, with cleavage along one side. Monoclinic.	**BCG**
Clinozoisite		White, grey or pale straw. ⎭	**BG**
Zoisite		White, grey or pale green. Symmetrical cross-sections. Orthorhombic.	**BG**

Olivine	7	Pale *olivine green*. Quartz-like, but with a tendency to alteration, to serpentine. Orthorhombic.	**AF**

Serpentine minerals	2–3	(Often contains embedded small grains of harder minerals.) Pale to dark *serpentine green*. Massive or fibrous. *Asbestos* will mat if crushed.	AF
Chlorite	1–3	*Chlorite green. One cleavage. Cleavage flakes bend, not elastic.*	ABCEFG
Talc	1	*White or pale* greenish. Feels *greasy. One cleavage. Cleavage flakes bend, but not elastic.*	AF
Micas	2½	*One cleavage. Cleavage flakes elastic* (springy). *Pseudohexagonal.*	
Biotite		Green–black or brown–black.	BCDEG
Phlogopite		Yellow, brown or green. In marble or ultramafic rock.	AF
White micas		White, grey or various pale colours (often greenish).	CDEFG
		Phengite variety. *Green*, occasionally brown.	H
'Sericite'	1–3	*Fine-grained*, white or green. Mica-like or talc-like minerals (including some white micas and phengites).	CDEFG
Gypsum	2	White or grey. One cleavage. Often fibrous. Simple twins.	F

10.2.2 'Metamorphic minerals', often as grains embedded in bulk minerals

Garnet group	6–7½	*Dodecahedral habit. No cleavage. Cubic.*	
'Pyralspite' garnets		Red, brown or purple. (Normal garnets.)	ABDEGH
Grossular		Pink–yellow. Rarely greenish.	G
Hydrogrossular		Pale pink. Often not as identifiable grains.	AB
Andradite		Yellow, brown or green. In iron-rich rocks and skarns.	

ALUMINOUS MINERALS OF SCHISTS:

Stilpnomelane	3–4	Deep brown or green–black, with one good cleavage. *Biotite-like*, but with *a poor second cleavage*, and *brittle cleavage flakes*. Thin blades or flakes, but more often in shapeless blobs and patches.	CEH
Chloritoid	6–7	Very dark green or green–black. One good cleavage. Somewhat *biotite-like*, but with *brittle cleavage flakes*. Equi-dimensional *tablet shapes*, or flaky aggregates. Pseudohexagonal.	E+G
Staurolite	7	*Brown prisms*, often flat-ended. Twins common. One poor cleavage. *Orthorhombic.*	E+

ALUMINOUS MINERALS OF SCHISTS AND GNEISSES:

Kyanite 5 along grain length *White* or *blue glassy blades.* DE E+
7 across the grain

Cleavages along and across the length, at 85°. Simple parallel twins. Triclinic.

Andalusite 7 Pink, or white or brown, squat or elongate CDE E+
prisms with *square cross-sections. Two
equivalent cleavages parallel to prism
faces. Orthorhombic.* Alters off-white
easily. Sometimes with inclusions in the
shape of a cross ('chiastolite').

Sillimanite 7 Whitish. Fibrous, or in needles or long CDE E+
prisms. Often as fibrous bundles. *One
cleavage along length,* on diagonal of
diamond shape of cross-sections.
Orthorhombic. (By comparison,
zoisite needles are shorter, with more
rounded-off ends and corners.)

Cordierite 7 White or blue, and *quartz-like, but* CE E+
tends to alter easily, around edges, in
concentric zones, or throughout. *In
schists, more often as ovoid inclusion-
filled porphyroblasts. Pseudohexagonal.*

CALC-ALUMINOUS MINERALS (see also plagioclase and epidote group).

Zeolite group 3–6 White. Massive, fibrous or blocky. Best BC
developed filling veins or cavities, but also
replaces *plagioclase* or *albite* in
otherwise little altered igneous rocks.
Belong to several crystal systems.

Prehnite 6 *Prehnite green.* Brittle tablets with one B
good cleavage. Occurrence: as zeolites.
Orthorhombic.

Pumpellyite 6 *Blue–green* masses and in veins. BH

Lawsonite 6 White, pale blue, yellow or pale orange. BGH
Equidimensional tablets or short prisms,
with square or diamond sections. Two best
cleavages (of many) at 90°. Orthorhombic.

MINERALS OF HIGH GRADE MARBLES (see also olivine, pyroxenes and garnets):

Wollastonite 5 White, pale grey or green. *Fibrous or* FG
splintery blades, with 3 cleavages along
length. (Tremolite-like.) Triclinic.

Monticellite 5½ Colourless grey. Poor cleavage. Often F
ovoid. Orthorhombic.

Melilite 5–6 White, grey–green or brown. Poor F
cleavages. Sometimes square-sectioned
prisms. *Tetragonal.*

| Periclase | 6 | Grey–white, yellow, brown or green–black. Spherical or octahedral grains. *Cleavage cubes. Easily altered to brucite.* *Cubic.* | AF |
| Brucite | 2½ | White or greenish–grey. One cleavage. Normally either as *streaks and masses in serpentine*, or *pseudomorphs after periclase in marble*. (Compared to talc: Less soft. Never found in contact with quartz.) | AF |

VERY HARD MINERALS:

Sapphirine	7½	Light blue or green. Cleavage poor. *Monoclinic.*	ACE
Spinel	8	Black or red. Octahedral grains. No cleavage. *Cubic.*	AFG
Topaz	8	Yellow or white prisms. One *cleavage, at 90° to grain length. Orthorhombic.* Mainly metasomatic, with fluorite and micas.	
Corundum	9	White or yellowish, or various colours. Twins. *Barrel-shaped grains. Trigonal.* In xenoliths and at very hot igneous contacts.	

OTHER METAMORPHIC MINERALS:

Scapolite	5–6	White, or various pale tints. Often cloudily altered. Squat or elongate *square-sectioned prisms. Tetragonal.* Can be *indistinguishable from andalusite*, but occurs only in rocks with some calcic component, where andalusite cannot.	BG
Idocrase	6–7	Brown or green to yellow. Squat *square-sectioned prisms*, or as compact masses. *Tetragonal.*	AG
Humite group	6	*Deep brown, orange or yellow. Chunks;* or as fibres amongst fibrous diopside or tremolite. Cleavages poor.	AF

10.2.3 'Accessory Minerals', bearing minor chemical components

Tourmaline	7	Black striated prisms, needles and fans. More glossy than amphiboles. Poor cleavage. *Trigonal.*	CDE
Apatite	5	Off-white. May cleave. *Hexagonal.* Ubiquitous.	**All.**
Sphene	5	*Grey–brown wedge-shaped* grains, or shapeless. Two cleavages. Simple twins. *Monoclinic.*	ABCDG
Rutile	6	Red–brown to black. *Tetragonal.* Pale to red-brown streak.	CDEH
Hematite	5–6	Red to black. Thin splinters are blood-red. *Trigonal.* Red–brown streak.	**All.**
Ilmenite	6	Black. Igneous relic. Alters to sphene and magnetite. No cleavage. *Trigonal.* Black or brown streak. *Weakly magnetic.*	ABC
Chromite	5½	Brown–black. Igneous relic. Cleaves. *Cubic.* Dark brown to black streak. Not magnetic.	A
Magnetite	6	Black. No cleavage. *Octahedral grains. Cubic.* Black streak. *Strongly magnetic.*	**All.**
Pyrite	6	Brass yellow. *Striated cubes, pyritohedra,* or rarely octahedra. *Cubic.* Greenish, grey–black or brown–black streak. (Insoluble in hydrochloric acid.)	**All.**
Chalcopyrite	4	Brass yellow, deeper than coexisting pyrite. Shapeless. Very rarely as cubes. *Pseudocubic.* Green–black streak.	**All.**
Pyrrhotite	4	Brownish bronze. *Pseudohexagonal.* Black streak. *Magnetic.* Soluble in hydrochloric acid (giving hydrogen sulphide).	ABE
Graphite	1–2	Grey–black. Feels *greasy.* Will mark paper with grey–black streak.	EFG

10.3 Mineral-based rock-type names

Rock-type Name	Main Minerals	Some Likely Minerals	Category (Section 10.1)
Amphibolite	Hornblende + plagioclase	Garnet, biotite, clinopyroxene, quartz, clinozoisite	B
Anorthosite	Plagioclase (calcic)	Garnet, pyroxenes, amphiboles	B
Blueschist	Any assemblage including blue amphibole	Chlorite, epidote, albite, phengite, carbonates, garnet, aegirine–jadeite, lawsonite	H
Eclogite	Omphacite + garnet	Quartz, amphiboles, carbonates, clinozoisite, micas, kyanite, olivine	H
Granulite	Any assemblage including ortho- and clino-pyroxene		
	Acid granulite:	Quartz, K-feldspar, amphibole	CE
	Basic granulite:	Plagioclase, olivine, garnet, amphibole, spinel	AB
Greenschist	Albite + epidote + either actinolite, or chlorite, or both	Quartz, biotite, garnet, carbonates, white mica	B
Quartzite	Quartz	Micas, feldspars, garnet, Al-silicates	D
Marble	Carbonate minerals	Numerous possibilities. Names of carbonates are not additional minerals.	FG
	Calcite marble: calcite		
	Dolomite marble: dolomite		
Micaschist	Micas (often with quartz or carbonate or both)	See Section 10.1, F and G	CDEG
		Albite, K-feldspar, chlorite, garnet, aluminous minerals, actinolite, graphite	
Peridotite	Olivine	Pyroxenes, amphiboles, garnet, spinel, plagioclase, serpentine, talc, chromite, magnetite	A
Serpentine or Serpentinite	Serpentine minerals	Talc, carbonates, actinolite, diopside, chlorite, quartz, olivine, brucite	A

For Sections 10.4 and 10.5 (Checklists) turn to Inside Back Cover

10.4 Checklist for recording textures and fabrics

1 *Keywords* to introduce a general impression of the rock:
  Any applicable *fabric or structure-dependent name* (e.g. hornfels, slate, schist, gneiss):
  Whether apparently *isotropic* or *anisotropic*:
  Any *compositional patchiness* (e.g. augen, ribbons, stripes, fine-banding):
  Any *fissility* (schistosity; or cleavage, slaty, spaced, fracture, pencil, etc.):
  Fabric type (slate, mineral, shape, lamination, crenulation, pressure solution stripe) and *symmetry*:
2 *Detailed sketches* of examples of textures and elements of structure if visible in the field.
3 *Statements of time relationships* which are evident in the (sketched) textures and structural elements.
4 *Statements of the grain texture or fine structure features to which fabrics correspond,* with mineral names where applicable.
5 *Any further characterization of fabrics* (e.g. strings of grains on a schistosity plane; or a stretching lineation on the cleavage of a slate).
6 *Orientations* of all composition and fabric planes, linear features and intersection directions.
7 *Structural relationships* between fabrics, compositional bands and patches, and deformation bands. Shear sense of shear-zones. Refraction. Ages.
8 *Sense of asymmetry* of all intersections of fabrics, other fabrics and banding, and the displacement sense of crenulations, etc. not yet noted.
9 *Correspondence of local fabrics* (their minerals, directions and structural relations) *to regional patterns* elucidated by mapping and structural synthesis.